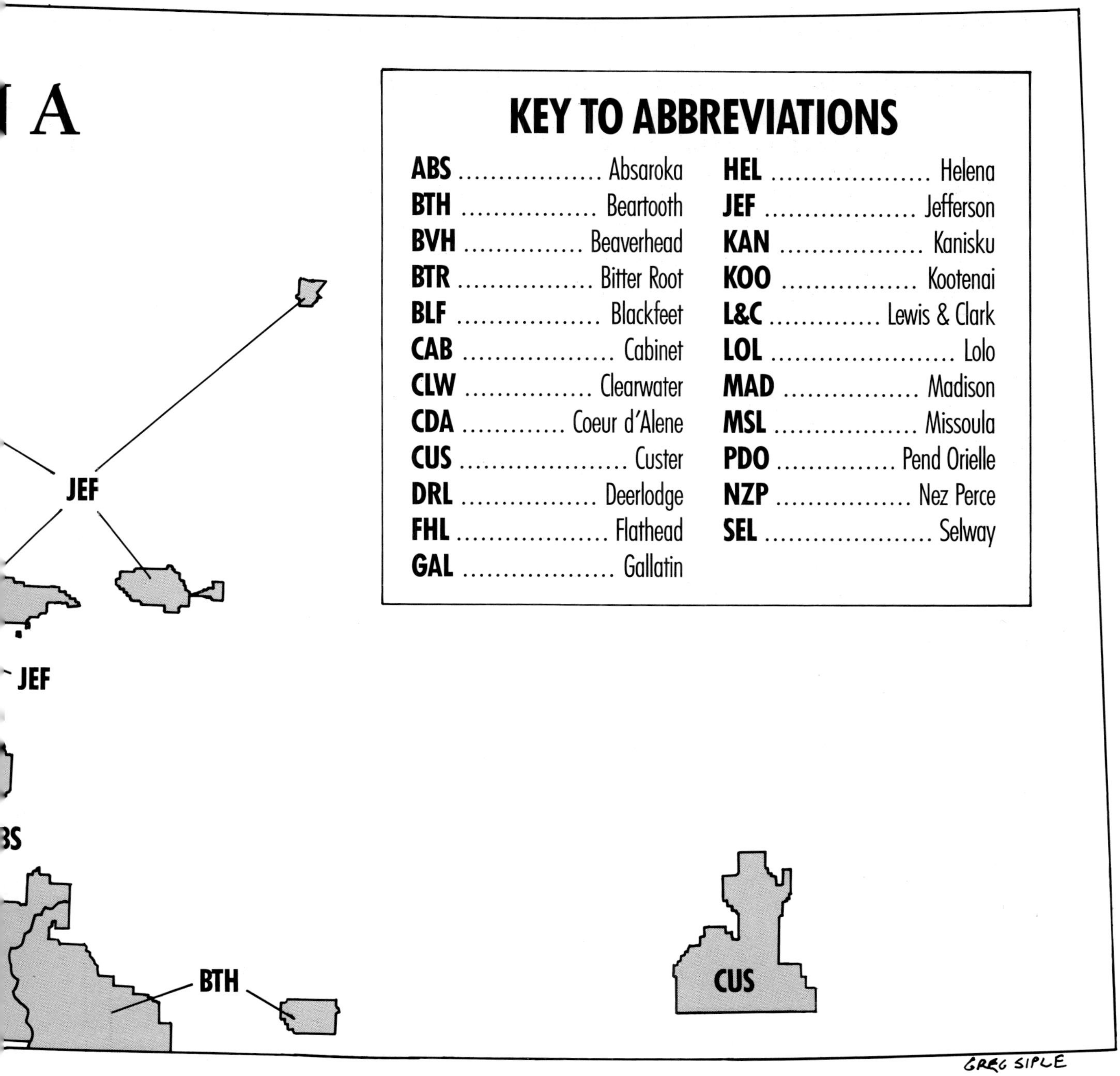

KEY TO ABBREVIATIONS

ABS	Absaroka	**HEL**	Helena
BTH	Beartooth	**JEF**	Jefferson
BVH	Beaverhead	**KAN**	Kanisku
BTR	Bitter Root	**KOO**	Kootenai
BLF	Blackfeet	**L&C**	Lewis & Clark
CAB	Cabinet	**LOL**	Lolo
CLW	Clearwater	**MAD**	Madison
CDA	Coeur d'Alene	**MSL**	Missoula
CUS	Custer	**PDO**	Pend Orielle
DRL	Deerlodge	**NZP**	Nez Perce
FHL	Flathead	**SEL**	Selway
GAL	Gallatin		

H.O. Bell Ford Agency in Missoula, Montana, delivered 50 ton-and-a-half trucks and 32 Ford pickups to the U.S. Forest Service in May, 1933. Owner H.O Bell poses with the trucks at Fort Missoula headquarters for the area's CCC program. COURTESY MRS. H.O. BELL

A Pictorial History of the U.S. Forest Service 1891-1945

NORTHERN REGION

by A. Richard Guth and Stan B. Cohen

NORTHERN REGION USFS
Striving for Excellence
CARING FOR THE LAND • SERVING PEOPLE

NATIONAL FORESTS
1891-1991

PICTORIAL HISTORIES PUBLISHING COMPANY
MISSOULA, MONTANA

LIBRARY OF CONGRESS
CATALOG CARD NUMBER 91-61013

ISBN 0-929521-36-6

First Printing: May 1991

Typography: Arrow Graphics, Missoula, Montana
Cover Art: Stan Hughes, Missoula, Montana
Cover Layout: Mike Egeler, Missoula, Montana

Photos in this book were provided by the U.S. Forest Service's National and Regional offices and various supervisors' offices and the Montana Historical Society (MHS). Other photos are credited accordingly.

PICTORIAL HISTORIES PUBLISHING CO., INC.
713 South Third Street West
Missoula, Montana 59801

Introduction

This is a pictorial history of the National Forests in the Northern Region (Region 1) and the employees of the U.S. Forest Service who worked in them. It includes the period from their beginning in 1891 when the first National Forest Reserves were created, through 1945, the end of World War II. It is a story about how some of the most magnificent lands in the world became public and their early management. It is a story of "conservation" in the United States and how it applied to the settling of the West. It displays the harsh physical demands required and the difficulties faced by pioneers in the Forest Service as they learned how to manage the National Forests. It permits an understanding of the elements that shaped the strong moral fiber that still permeates the U.S. Forest Service today.

The history of conservation in the United States begins shortly after the Civil War. At this time, a handful of influential scientists became alarmed at the rapid rate the forests of the United States were disappearing. When the first Europeans settled here, a blanket of trees covered much of the country. There was concern that removal of the forests, especially in drier areas, could result in the creation of new deserts.

The country's early social and economic structure, however, was largely dependent on agriculture—and you could not grow corn or cotton under a stand of trees. The trees had to go! In addition, following the Louisiana purchase, the government of the United States was very wealthy in land, but not money. According to many, the success of the democracy experiment in the United States depended on getting this land out of the government and into private ownership. Laws were passed to sell land cheap—or even give it away.

There were Federal grants to railroads and canal constructors; States were given land to sell or lease to help defray costs of education; settlers made claim to land under one of several laws—in total, by 1900, over a billion acres of land were taken out of the public domain to become private.

In the arid West, however, something else had to be done. If not, the forests would disappear, and with them, first the soil and then the water. Without these elements, there could not be any agriculture. Soil and water were basic building blocks needed for establishing civilization. The time was right to conserve these precious resources—it was time to let the principals of sound CONSERVATION—wise use—dictate the management of the western public forest lands. This was the beginning.

Preface

It is very appropriate to have this pictorial history of the Northern Region of the Forest Service published during this, the centennial year of our National Forests. The most significant feeling I had as I paged through this document was one of pride in the men and women who located, shaped and nurtured the splendid National Forests of the Northern Region we all enjoy today. They were magnificent and dedicated men and women. They experienced extremely hard work, long hours, low pay, poor living conditions, insurmountable odds—and thrived on these. There's was a challenge to stop the very destructive wild fires that periodically swept across the land. Fires that made human habitation risky or even impossible. It was also their responsibility to start managing the lands set aside for National Forests and demonstrate how they were of great value to all the people of the United States. They were strong believers in the conservation of our natural resources. They shared this concept with everyone who would listen to them. They practiced conservation every day of their lives. Read their words in this book. I know you will agree they had exciting times, and they were very serious about their responsibilities. Humor, however, was also important. They did not have television and were often isolated in remote locations. Spinning tall tales next to an evening fire became an art form. Practical jokes were a way of life, but also a demonstration of respect. Yes, we can be proud of them.

I am equally proud of the men and women presently serving in the Northern Region of the U.S.D.A. Forest Service. And I am confident their history will be as great as those who are a part of this book.

John Mumma
Northern Regional Forester
U.S.D.A. Forest Service

Table of Contents

BRITISH P
FLATHEAD FOREST
BLACKFEET INDIAN RES.
FLATHEAD
TETON
CHOTEAU
LEWIS AND CLARKE FOREST RESERVE
FLATHEAD INDIAN RES.
MISSOULA
LEWIS AND CLARKE
CASCADE
GREAT FALLS
FT. BENTON
KALISPELL
MISSOULA
POWELL
GRANITE
PHILLIPSBURG
HAMILTON
DEER LODGE
ANACONDA
HELENA
JEFFERSON
BROADWATER
TOWNSEND
MEAGHER
WHITE SULPHUR SPS.
SILVER BOW
BEAVERHEAD
DILLON
MADISON
VIRGINIA CITY
GALLATIN
BOZEMAN
PARK
LIVINGSTON
BITTER ROOT MOUNTAINS
BITTER ROOT
IDAHO
ROCKY MOUNTAINS
LEMHI IND. RES.
YELLOWSTONE NATIONAL PARK
YELLOWSTONE LAKE
3d Stand. Par. N. Boise Mer.
LEGEND.
SURVEYOR GENERALS OFFICE.
U.S. LAND OFFICES.
MILITARY RESERVATIONS.
INDIAN
FOREST RESERVATIONS AND YELLOWSTONE NATIONAL PARK.
TOWNSHIPS NOT SUBDIVIDED.
GALLATIN FOREST RES.
I hereby certify that I have examined the above map and found it in accordance with the provisions of House Bill No. 138 passed by the Seventh Legislature of Montana, and I hereby approve the same.
J.H. Calderhead
Montana State Auditor
John C. Van Hook
Copyright applied for.

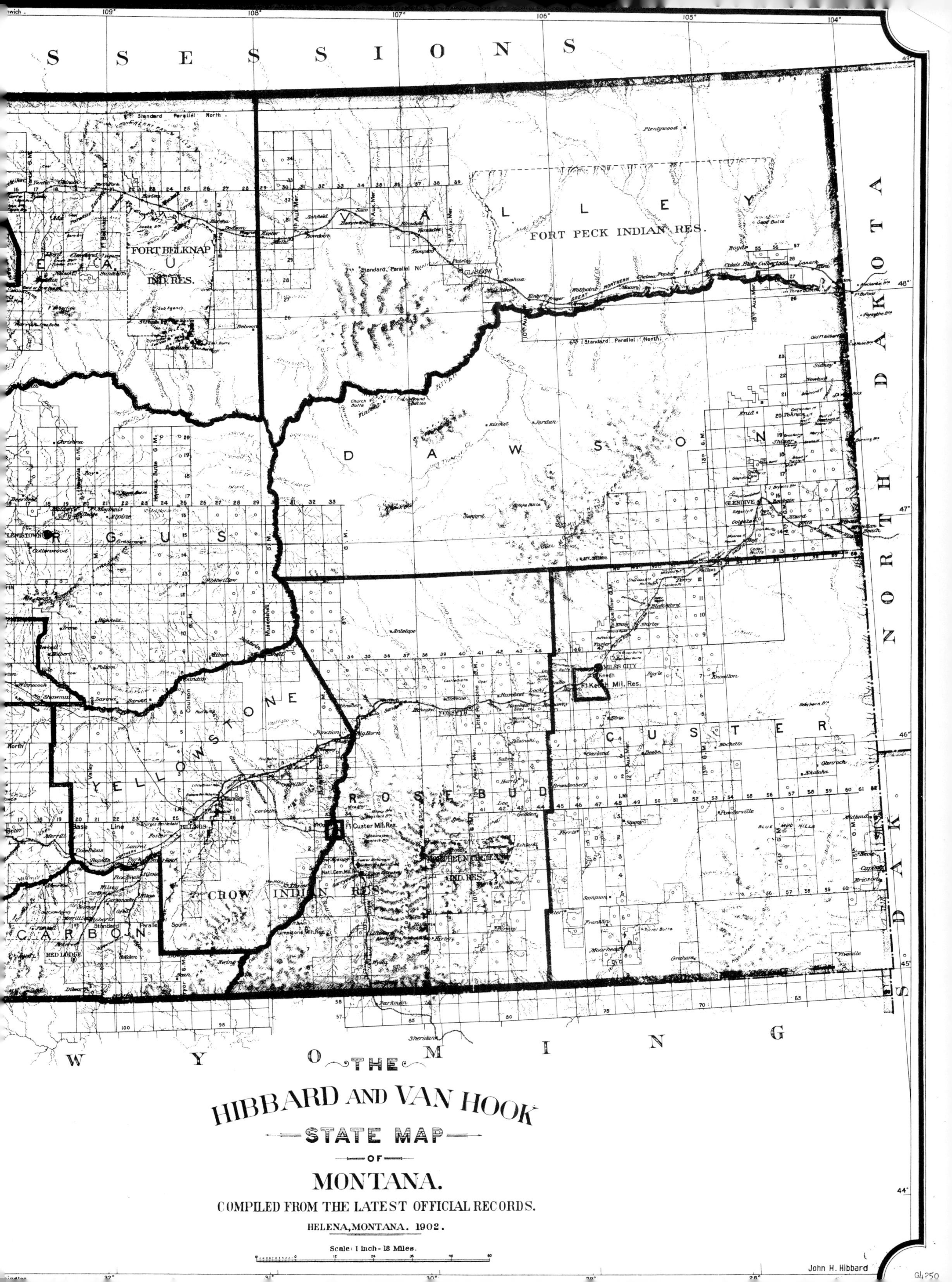

THE
HIBBARD AND VAN HOOK
STATE MAP
OF
MONTANA.
COMPILED FROM THE LATEST OFFICIAL RECORDS.
HELENA, MONTANA. 1902.
Scale: 1 inch - 18 Miles.
John H. Hibbard
POSSESSIONS
VALLEY
FORT PECK INDIAN RES.
FORT BELKNAP IND. RES.
DAWSON
FERGUS
YELLOWSTONE
ROSEBUD
CUSTER
CARBON
CROW INDIAN RES.
NORTHERN CHEYENNE IND. RES.
Ft. Custer Mil. Res.
Ft. Keogh Mil. Res.
NORTH DAKOTA
SOUTH DAKOTA
WYOMING
GLASGOW
GLENDIVE
MILES CITY
LEWISTOWN
BILLINGS
FORSYTH
RED LODGE
Plentywood
Sheridan

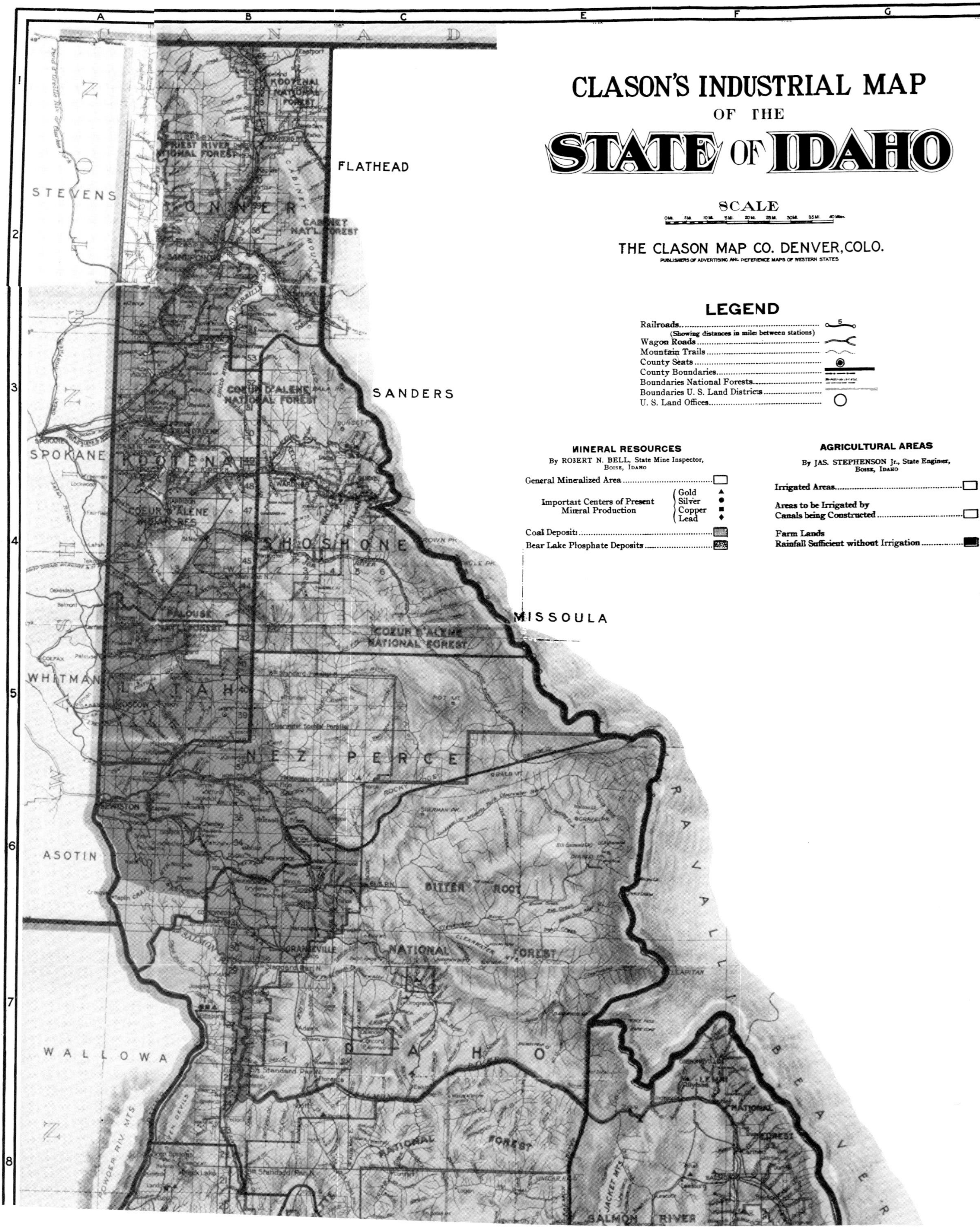
CLASON'S INDUSTRIAL MAP
OF THE
STATE OF IDAHO
SCALE
THE CLASON MAP CO. DENVER, COLO.
PUBLISHERS OF ADVERTISING AND REFERENCE MAPS OF WESTERN STATES
LEGEND
Railroads (Showing distances in miles between stations)
Wagon Roads
Mountain Trails
County Seats
County Boundaries
Boundaries National Forests
Boundaries U. S. Land Districts
U. S. Land Offices
MINERAL RESOURCES
By ROBERT N. BELL, State Mine Inspector, Boise, Idaho
General Mineralized Area
Important Centers of Present Mineral Production
Gold
Silver
Copper
Lead
Coal Deposits
Bear Lake Phosphate Deposits
AGRICULTURAL AREAS
By JAS. STEPHENSON Jr., State Engineer, Boise, Idaho
Irrigated Areas
Areas to be Irrigated by Canals being Constructed
Farm Lands Rainfall Sufficient without Irrigation
CANADA
FLATHEAD
STEVENS
BONNER
SANDERS
SPOKANE
KOOTENAI
SHOSHONE
MISSOULA
WHITMAN
LATAH
NEZ PERCE
ASOTIN
WALLOWA
IDAHO
RAVALLI
BEAVERHEAD
LEMHI
SALMON RIVER
BITTER ROOT
NATIONAL FOREST
COEUR D'ALENE NATIONAL FOREST
PRIEST RIVER NATIONAL FOREST
KOOTENAI NATIONAL FOREST
CABINET NAT'L FOREST
PALOUSE NAT'L FOREST
COEUR D'ALENE INDIAN RES.
LEWISTON
MOSCOW
GRANGEVILLE
SANDPOINT

CHAPTER ONE: Timber

A stand of western larch in the Kootenai National Forest in 1921.

Timber

Logs from trees and lumber from logs were needed for the construction of homes and buildings. Wood was also used for fuel, charcoal, paper pulp, furniture and countless other uses. And the National Forests of the Northern Region had millions of acres covered with stands of trees. These stands, managed carefully, provided an unending supply of wood.

Before the white settlers, Native Americans, living in what is now the Northern Region, were dependent on wood. They used it for cooking, heating, and drying food as well as making their bows, arrows and other tools. Pine poles supported their tepees. These trees still bear a name describing this purpose—lodgepole pine.

When white settlers came, they brought tools for working wood with them—axes, saws, planes, drills, chisels, hammers and more. These were prized possessions for shaping wood. In 1848 one of the first sawmills for producing lumber in the Northern Region was set up at St. Mary's Mission in the Bitterroot Valley in Montana. It was a "whipsaw" mill. This is a mill where one person stands in a pit below a log and another stands on top of the log. They alternately pull a saw through the log cutting boards. The saw used in the St. Mary's Mission mill was made from the rim of a wagon wheel.

The discovery of gold and other precious metals brought a rush of pioneers to the Northern Region. These pioneers needed wood products for their developments, homes and businesses. Sawmills soon became part of these early communities, filling demands for construction lumber.

Early sawmills driven by water power were often established in conjunction with grist mills. An example would be the mill established in western Montana in 1864 by David Pattee at a trading post called Missoula Mills. Because of poor roads, it was usually easier to build a new sawmill near a new community than to haul lumber very far.

By the late 1880s, however, the expansion of mining ventures, the advent of railroads and the building of more new settlements expanded demands for wood products, including timbers for supporting underground mine shafts and for charcoal for smelting ore. The Anaconda Company alone, in Butte, Montana, was using 40,000 board feet of timber every day. In addition, they annually turned 300,000 cords of wood into charcoal. The railroads were building hundreds of miles of new spur lines and needed 3,000 ties per mile to support their new tracks. Still more trees were sawed into timbers and used by railroads to construct bridges.

Lumber was in high demand by settlers and other new residents for constructing homes, offices, stores and other buildings. It was a principal source of fuel for heat. Wood was very important. It was also very abundant. Before the Forest Reserves were created, most of the trees used for wood products were simply removed from the public domain without permission—sometimes after bribing an agent from the General Land Office. But who cared, the forests marching up every mountain appeared to be endless—they could never be used up.

Then, in 1891, when the Forest Reserves were created. Those who had been taking "free" timber from the public domain, were not at all pleased with this new concept. For many, this killed the goose that laid their golden eggs. Most of the people living in the West at that time, however, believed removal of forest lands from the public domain would stifle settlement, growth and prosperity. In 1905, when the U.S. Forest Service was created, there was no question that inexpensive wood products were still very essential to early settlement and development of the Region.

Before the creation of the U.S. Forest Service, just taking timber from the lands owned by the United States was a common practice. This is the McCune trespass area in what is now the Deerlodge National Forest. Between 1884 and 1892 the trees here were cut and the wood piled for use in the smelter at Anaconda, Montana. This picture was taken in 1923.

At the turn of the century, however, there were few, if any, who could conceive that the day would come when wood products from the National Forests of the Northern Region would be important in world trade. What would the reaction have been in that time to someone who predicted that log homes would someday be prefabricated in the Northern Region for shipment to Japan? The most advanced thinkers in 1900 believed the western forests should be protected and used wisely to support local commerce and industry. There was no need to consider a timber industry beyond this need.

A 1907 U.S. Department of Agriculture publication, "The Use Of The National Forests" provides the following information about timber:

> "What happens to the timber and wood? The timber and stone act does not apply to the National Forests. The title to lands chiefly for timber cannot pass from the government.
>
> "But are the timber and wood locked up? Very far from it. The timber is there to be used now and in the future. It is given away, for domestic use, to the man with a home and to the prospector developing his claim. They get it for the asking, free of charge. When wanted for commercial purposes, timber is sold to the small man and the big man—sold promptly and at a reasonable cost. The small man can buy a few thousand feet; the

big man can buy many million feet, provided it is a good thing for all the people to let him purchase a large amount, but not otherwise. The local demand is always considered first. There is no chance for monopoly, because the Secretary of Agriculture must by law sell as much or as little as he thinks best, to whom and at whatever price he thinks will best serve the interests of all the people.

"Thus the timber is there, first of all to be used. The more it is used the better. Far from being locked up, it is, on the contrary, opened up, and opened up on fair terms to all alike. When it is on the open public domain it is often very hard and sometimes impossible for the small man to get it and hold it, because he is shoved aside by the big timber corporations with which he can not compete. On National Forests the Government holds the timber with a special view to its use by the small user. At the present time nine-tenths of the timber sales on National Forests are for less than $500.

"In 1906 75,000,000 board feet of timber were given away and 700,000,000 board feet were sold.

"The National Forests occupy high mountain lands, rough and rocky, and which will always be of value chiefly for the production of timber and wood."

Sometimes logs being floated down river did not behave and there were tremendous log jams. Dynamite was usually used to break up the logs forming the jam. It took experienced River Hogs to identify the "key" logs, set the dynamite and get the logs moving again. Here is a log jam at Bonner, Montana in 1899.

A big log jam in 1908.

Strip cutting in lodgepole pine, Deerlodge National Forest, 1908. The strips with standing trees are supposed to serve as a seed source to reproduce the cut strips.

Block cutting in lodgepole pine in 1906. All trees are cut except for scattered groups or blocks that are 75 feet square. This was the Allen Co. Timber Sale in the Deerlodge National Forest. The purpose for leaving the blocks is to let them serve as seed sources to restock cut areas.

Mature white pine like this was "bread and butter" for the early timber industry in the Northern Region of the U.S. Forest Service.

Also, "The timber is for use. The cuttings do not damage the Forest because the lumbering operations are so carefully done that the stand is left in first-class condition for a second crop, and after the third crop and any number of future crops."

Concerning timber harvest on private lands, this publication states:

> "The actual results on private lands where the owners do not care after they have skinned them are quite different. These lands are usually cut over with the sole object of getting everything possible out of them in one stroke. They are stripped of timber while the slashings which are left on the ground make good fire traps. Very soon the whole area burns over and the ground becomes a nonproductive waste. A glance from a car window in Michigan, Wisconsin or Minnesota shows the now absolutely ruined lands which but a short time ago produced magnificent stands of white pine. Think of the great wealth which the people of these States might have made permanent, simply by using the Forests wisely.
>
> "Then, again, wood is so very essential in everyday life that it seems unwise to let it be monopolized by individuals or corporations. Actual results show that when public timber lands pass out of the government's hands they eventually, and often very quickly, fall into the hands of the big concerns, which rarely show the slightest tendency to handle them for the greatest good of the people in the long run.
>
> "On the National Forest the present and future local demand is always considered first. The government tries to see that there shall always be enough timber and wood on hand for use by the home builder, the prospector, the miner, the small mill man, the stockman and all kinds of local industries. If local needs promise to consume it all, nothing is allowed to be shipped out of the country. If it were in the hands of individual or corporate owners, it would very likely be shipped out, regardless of local needs. It would seek the best market. If it were sold locally, the users would have to pay whatever price the owner might demand and the price might be unfair."

Because of careless logging and devastation occurring on private lands, early Forest Service employees, from top to bottom, were determined that the harvest of timber from the National Forests would only be done according to sound principles of conservation.

Finally, there is a "How to do it" section that says:

> "Free use–To get the free use of timber and wood, ask the nearest Ranger for a permit. It is given away free to settlers, farmers, prospectors and others for domestic use, and to school and road districts. It is not given away free for any kind of commercial use.
>
> "Do not cut or remove the timber or wood until a permit is obtained. In great emergencies, when you can not reach a Forest officer and when serious loss would result by not waiting for one, take what is actually needed with as little damage as possible and report to the Ranger afterwards."

District Rangers were given authority to sell up to $50 worth of timber. Over $50 but under $500 worth of timber could be sold by the Forest Supervisor. Over $500 required advice from Washington. For those interested in buying timber, the publication says:

> "Under any circumstances the papers must always pass through the Supervisor's hands. He is the man to deal with. The purchaser should first locate the timber he wants to buy. After it has been cruised the Supervisor will agree with the purchaser upon the terms of the sale. Cutting may begin as soon as the timber is marked for removal and payment has been made as called for.
>
> "Do not begin to cut or remove timber on any sale until permission to do so has been granted by a Forest officer."

From the start of the U.S. Forest Service in 1905 and through World War II, most of the timber in the National Forests of the Northern Region was not economical to harvest and process for national markets. Small sawmills, process-

A stand of western larch in the Flathead National Forest in 1908. In the early years, because it was pitchy and tended to have a twisted or spiral grain, larch was not one of the more sought after species.

Ponderosa pine stand (also called yellow pine) in the Bitterroot National Forest. Ponderosa is a "hard" pine and not as easy to work with hand carpenter tools as the "soft" white pine. The primary reason for the parklike appearance is a result of frequent, natural fires killing the underbrush but which are not hot enough to do more than scorch the thick, corky bark of the large ponderosa pine trees.

Railroad lines for logging began to criss-cross the timber stands, mostly on the private lands, of the Northern Region. This is a line owned by the Clearwater Timber Co. in 1928 near Pierce, Idaho in the Clearwater National Forest.

ing lumber for local use were common throughout the region. There was, however, as mentioned above, a hefty demand in wood for mines and railroads. In fact, perhaps the largest clearcut in the entire National Forest System happened adjacent to Butte, Montana. The wood was used in the development of the extensive underground mines near this city. The other exceptional areas that commanded national attention were the western Montana and north Idaho stands of western white pine.

By 1905, the eastern white pine stands were about gone in the area east of the Mississippi River. Now big timber industries were moving in on the magnificent stands of western white pine and cedar in western Montana and north Idaho. Pinchot's warning about large timber industries threatening timber supplies was not heard by very many in the West. Big sawmills meant big cities.

In 1905 there were few among the new, young, U.S. Forest Service employees that lost any sleep over thoughts of a timber famine. Their job, according to "The Use Of The National Forests," was to provide timber to meet local needs. This included the timber industry and their sawmills. Besides, there were not that many settlers in or near the western Montana and north Idaho National Forests who needed lumber for their developments.

In 1954, in the timber sale files of the Northern Region, the Dalken Lumber Co. Sale of 8/20/12 had an undated note. The note read "The largest sale in the (Northern) Region." The contract was approved September 9, 1913 by Henry S. Graves, then Chief of the U.S. Forest Service. The sale covered more than 18,000 acres of National Forest land in north Idaho. It was finally closed about 1930.

The sale was estimated to contain about 240 million board feet of logs, about half of which was white pine, and almost 200,000 cedar poles. The bid price for white pine was $5.00 per thousand board feet, yellow pine was $2.00 per thousand board feet, spruce and dead white pine were $1.00 per thousand board feet and all other species were 50 cents per thousand board feet. This was hardly a timber sale for a local settler's building material.

Small sales, however, were frequently made. According to I.V. Anderson, in 1922 he and Howard Larson, District Ranger at Plains, Montana were ". . . cruising a small jag of timber for a stump farmer in the vicinity of Paradise, Montana. We had plenty of applicants for small jags of timber around 200,000

There were sawmills around when the Forest Service began. This is Hall's Sawmill in 1907. It was operating under a permit with the old Sioux National Forest in Custer County, Montana, in connection with a timber sale. The mill has recently burned out—note, there is no spark arrestor on the smoke stack.

"Scaling" means determining how many board feet (1 inch by 1 inch by 1 foot) can be cut from a log in a competent sawmill. If ever there was a situation more primed for controversy between buyers and sellers than scaling logs, it is still coming in the future. The length of the log, straightness, presence of knot or rot, average diameter inside the bark, are all elements that must be considered. These scalers are at work in the Lolo National Forest in 1908.

(board) feet. It made a good-sized job for a farmer to make a stake of $1500-$2000. With sawlogs selling for from $8 to $12 per thousand (board) feet at the river (Clarks Fork), none of them got rich at logging."

Another typical big industry offering was the 1926 Kalispell Creek Sale in the Kaniksu National Forest. It was purchased by the Diamond Match Company. According to Henry A. Peterson:

"In those days sales were for white pine. Hemlock was included for silvicultural reasons and usually appraised for a minus value which was carried by the white pine values." The white pine stumpage was reduced to offset the loss in the appraised price of the hemlock.

"They had two logging trains consisting of a locomotive and nine log cars each. The locomotives were 20-ton, gasoline powered Plymouths with extra ballast. The engineer shifted gears more or less like a truck." The logs were dumped into Priest Lake. and in the Spring, driven down the river to the sawmill at Cusick, Washington. Concerning the hemlock in this sale, most was not merchantable. To be merchantable in 1926, a hemlock tree had to be 50 percent sound as sawlogs. The Kaniksu National Forest Supervisor at that time, Karl Klehm, "was a proponent of clear cut and burn of such climax types."

Most of the skidding in this sale was done by horses. The only exceptions were a few long, flat areas that were skidded by tractor. There were very few old-time lumberjacks on the job. Most of the workers were natives from around Priest River who went home on weekends.

"Prior to World War I," according to Peterson, "I worked in several old-time logging camps. Generally they all served very good food, but living conditions were terrible! There were lumberjacks in the camps with colorful names such as Wood-em-up George Yarno and Moonlight Joe Piatt." Peterson continues, "The old time lumberjack is a vanishing breed of man; very few are alive today (About 1956). They are the men who worked as lumberjacks in the Lake States during their younger days and migrated to the West as the timber diminished in the East. They were of many nationalities with Scandinavians and French-Canadians well represented. Prior to World War I they carried their own bedroll and each had a packsack with all of his personal belongings, and quite often his entire fortune.

"The bunkhouses usually were large and constructed of logs. A row of double decked wooden bunks was along one side. A double drum style heating

How much rectangular lumber should be cut in a sawmill from a round log? It's the scaler's job to determine this. The logging contractor pays the United States so much per thousand board feet (MBF) based on the scaler's figures. This amount is called "stumpage."

A double ended Clyde Loader working near Libby and the Kootenai National Forest.

Peaveys, a long pole with a spike and hook at one end, were the basic tool used by the log drivers, called "River Hogs," to keep the logs moving downstream. They had to be cat-footed and wear caulked boots to successfully move across the floating logs. Here is a crew working on a jam in the Gallatin River, Gallatin National Forest. Note this jam consists of hand hewed railroad ties.

stove made from oil barrels was located in the center of the building between the rows of bunks. There was a 2×12 plank (deacon) seat attached to the lower bunk and extending the full length of the building except in one corner which was occupied by a wooden trough on an incline that was used for washup. There were no drying rooms or facilities for taking a bath or doing laundry. However, they did have one luxury—hot water with which to wash. A hot water tank was an open metal barrel near the heating stove. The water was heated with pipe coils in the stove.

"Most of the lumberjacks wore woolen two-piece underwear winter and summer. Usually they came in sweaty and often wet and hung their wet sweaty clothes and socks around the heating stove to dry. The smell was indescribable. Talk about air pollution and smog! This was something, with tobacco smoke, and 50 men drying their clothes and socks, and all the other odors of hard working men who seldom took a bath or changed their clothes for weeks and sometimes months at a time for the simple reason that there were no facilities for bathing or laundering.

"There was one redeeming feature of the old time logging camp; they had good cooks, served good food, and plenty of it. The old salt pork and bean days of the past were long gone. Those camps were home to many of those fellows. They were single and footloose. Many stayed over winter and some were short-staked and moved from camp to camp. It was a policy that if they came to the camp looking for a job they received two or three free meals and lodging if there were empty bunks, otherwise they slept on the benches. They had a code of ethics. They would never ask for a job from the foreman; it was up to the foreman to offer them a job. Most were specialists, i.e., teamster, sawyer, swamper, canthook man, etc., if there wasn't an opening for his line of work, he took off for the next camp. The cookhouse also had a code of ethics. No talking at the table except to request passing of food. Should a greenhorn or someone start visiting, the cook or second cook, armed with a large butcher knife, would tap you on the shoulder and remind you to eat and get out."

Concerning the results of the Kalispell Creek Sale, Peterson writes, "It is believed that the Kalispell Creek drainage, as a whole, especially in the State of Washington, experienced the most complete and drastic face lifting of any comparable drainage within any Forest in Region One. Personally I saw the drainage prior to a tree being cut. When I visited the area during the late years, it is difficult to realize or visualize the changes that had transformed the old to the present appearance."

According to C.B. Sutliff: "The first logging trucks in the timber industry were equipped with solid rubber tires. Because they were hard on the normal dirt roads in the area, much of the hauling was done after the ground froze in the winter. 'Pole roads,' made of hewed timbers, were often constructed over unstable or wet areas.

"The pole roads were constructed by laying hewed logs, 12 to 16 inches in diameter, side by side with the hewed side up. A pole, about 6 to 8 inches in diameter was spiked to the inside of the inner hewed log to serve as a guard rail.

"Meeting a loaded truck while driving a car on this road was terrifying since backing up quickly around curves was very difficult. The best method of negotiating such roads was to pull off the road at every turnout. Then listen carefully, proceed and hope for the best."

Each lumber company usually had at least one hewed log construction crew. Many were of Swedish extraction. Usually one of the crew was an expert with a broad axe. Hemlock, if available, was used in this construction.

In 1912, Mark Watkins was looking for work in the logging camps that had been established to salvage timber killed in the 1910 fire. It was a 20-mile hike along a primitive tote road. He was hired as a "bull cook" by "Slick Jack" Cox at the first camp he came to. His brother was hired as a "chute greaser."

Timber was being harvested on the south side of the Little North Fork of the Coeur d'Alene River. According to Watkins: "The easiest and most economical way to get the logs down to the river was by building chutes down the mountainside in which to slide the logs by letting gravity do the work. These chutes were made by mounting two logs side by side on cross timbers. The proper slope was maintained by building cribbing under the low places; the inner side of the chute logs were hewed into a "V" shape with a broadax. To prevent the logs from jumping out of the chute on the curves, the outer side of curves were built up two or three logs high. It was necessary to grease the sides of the chute, especially on the curves, to keep the friction from slowing down the logs' speed. Also, the chute would last longer if it was greased. The grease was applied with a wooden paddle. The job was not difficult, but the greaser had to keep alert as the logs traveled at high speed and would sometimes jump the chute at the curves."

In 1909 John Preston had a job looking after the timber sales in the National Forests of eastern Montana. He was responsible for silvicultural systems, mark ing standards, running out sale boundaries, appraising the stumpage and drawing up sale contracts.

During the winter of 1909 Preston visited a timber sale in the Crazy Mountains at the request of V. Giffert Lantry, Supervisor of the Absaroka National Forest. Preston traveled to Big Timber, took a stage coach as far as he could, then walked 12 miles to the mouth of Cottonwood Creek. He met the Ranger and rode another six miles to the timber sale area.

According to Preston, "The Supervisor had been making $100 sales to local sawmill men operating under the name of Mosback and Eicke. The sawmill was a little one-horse portable affair which could cut, when pushed, perhaps 3,000 feet in ten hours. The camp was one of those combinations where one-half of the building was devoted to housing of the horses and the other half a combination cookhouse and bunkhouse. The door, I remember, was so small that one had to bend almost double before entering."

In the early days, the fastest, easiest, and frequently the only way to get logs from the forest to the sawmill was to take advantage of gravity and water. Good highways and big trucks would come later.

What Mosbeck and Eicke wanted was an advertised sale. The Supervisor, however, did not have any Rangers "sufficiently experienced to handle even so simple a proposition." Preston layed out the sale boundaries, marked the timber, made out the sale application and returned to Livingston. Here he helped the Supervisor fix up the necessary papers for advertisement, and arranged for 'advance cutting' since logging must proceed. According to Preston, "All of this expedition to the sale area and return to Livingston required something like 10 days." Preston said, "The Rangers of those days knew only a little about what forestry was all about, but they were anxious to learn."

C.S. Webb discussed salvage operations following the 1910 fires. According to Webb, "The 1910 fires burned large areas in the St. Joe, Coeur d'Alene and Cabinet National Forests. By now logging was brisk in these areas. The Wintons built a mill at Rose Lake, Idaho, and started cutting on Independence Creek in 1911, driving the logs all the way to Rose Lake. Herrick built the Milwaukee Lumber Company Mill at St. Maries in 1911, and St. Maries Lumber Company, a Michigan concern—built on there in 1912." Webb indicated there were more new mills in the area.

He then described the timber, "Timber in the white pine belt swept by the 1910 fire was mostly large old growth, quite sound, choice white pine. Logging was somewhat difficult, with big horses furnishing the skidding power. Horse-skidding, chutes, flumes, railroads and river-driving were the means of transportation. There were no "cats", no trucks. Sleigh haul was employed a lot on Priest River, Pack River and the Kootenai. Big river drives were the order of the day each spring on the Kootenai, Pack, Priest, Couer d'Alene and St. Joe Rivers. The "river pigs" (men hired to move the logs down the rivers) had considerable leeway in moving about."

By 1913, Webb indicates, ". . . deterioration of sapwood had rendered most of it unmerchantable. The timber over large areas had blown flat on the ground under pressure of the heavy winds ahead of the 1910 holocaust. The timber therefore was not burned, nor checked to any appreciable extent. Some sapwood had blued but was firm. But, most of the logs were scaled inside the sapwood. Nothing smaller than about 14 inches at the top end was removed. We scaled out many trainloads of short logs which, even under these conditions, averaged 6 logs per thousand (board) feet. It was choice white pine, running a high percent of clear, selects and shop. Stumpage price was $2 per thousand board feet. But this bonanza was short-lived. By the end of 1915, the removal of 1910 burned timber, except for cedar, came to an end."

"Koch Specials" were white pine trees left uncut in a harvest area. They were left as a seed source to start a new stand of white pine in the areas.

According to Sam Billings, in 1924 tamarack (larch) and Douglas fir hand-hewed railroad ties were being bought by the Great Northern in Eureka, Montana for 43 cents each.

Billings was a member of the 1926 party cruising timber in the Kaniksu National Forest, which had burned that year. This area included the Kalispell Creek drainage and the Diamond Match Company's 1926 Kalispell Creek Sale described previously by Henry A. Peterson. He says, "There was so much acreage burned on that forest (Kaniksu) that year (1926) they called it 'The Lucky Strike Forest' after the cigarette of that name (it was toasted). Some of the finest Idaho white pine stands that ever grew were burned that year in Harvey, Grave, and Kalispell Creeks. And I guess, because of a lack of a road system in those drainages, but little of it was salvaged."

He also wrote, "During the fall of 1926 and 1927 I helped mark considerable timber on quite a number of timber sales. The marking policy in mature white pine stands was to leave three to five full crowned trees per acre that are at least 24 inches in diameter. These were called 'Koch Specials.' The balance of the trees were marked for cutting. The 'Koch specials' were named after Elers Koch, for years Assistant Regional Forester, in charge of timber management."

Elers Koch, who spent 22 years, 1921 to 1943, in the Northern Region as Assistant Regional Forester, Chief of the Division of Timber Management, was very familiar with the Region and all of its timber types. These included the lodgepole pine forests of eastern Montana; the ponderosa pine on dryer sites; the Douglas fir and larch stands in western Montana; and the white pine and cedar stands of north Idaho.

According to Koch, "The western white pine has always been our most valuable species. Often when white pine stumpage is eagerly in demand from $5 to S6 a thousand feet, the associate species, Douglas fir, larch, white fir and hemlock, would show a very low or even a minus value." This complicated things ". . . since our objective after harvesting a crop of mature timber is to reproduce as much as possible of the valuable white pine. Much of our activity and study of the forestry problems of the district centered about working our practicable methods of handling the white pine type."

Koch, venting his emotions, described the original white pine forests and what was happening to them. "There is no more beautiful forest than a mature white pine stand, with the noble gray ridged columns of the white pine rising

The Big Trees

A tree that can grow into a real giant is the western white cedar found in wet areas and stream bottoms in the western part of the Northern Region. Rot resistant, cedar is valuable for shingles or other uses were wood frequently gets wet. Large trees like this are usually hollow and very dangerous to cut.

Using a "misery whip" near Rodman, Montana in the Flathead National Forest. The tree is a Douglas fir, another important lumber producing species in the Northern Region.

Spruce, seldom found in pure stands, can grow into large trees in the western part of the Northern Region. This giant was cut near the head of Bristow Creek in the Kootenai National Forest.

There were two stages in felling a tree. First a notch was cut with a sharp axe. For experienced fallers, this determined the direction a tree would fall.

Next the fallers went to work with a two-man cross-cut saw, also known as a "misery whip." These were the principal tools used to cut trees during the period covered by this history. Teamwork was essential.

Logs piled on skidways at the end of a chute in the Lolo National Forest.

Scaling logs as they were rolled off railroad cars gave the scaler a good look at the logs and permitted a more accurate scale.

Scaling decked logs required more than average skill and agility. Holding his scale stick in his right hand, his scale book in his left, the hammer he uses to stamp logs after he scales them is in his belt, and his pencil is behind his ear. If the logs in deck start rolling, he'll jump for his life! This scaler is scaling logs in 1924 from the Skookum Creek Sale, Coeur d'Alene National Forest.

Getting the logs off the mountain was not always the end of their trip. Here, in 1923, logs are being loaded onto sleighs at the end of a chute in the Beardmore Timber Sale Area, Kaniksu National Forest.

A team of horses skidding logs to a landing on the Kaniksu National Forest in 1923.

above the vivid green crowns of the associated hemlock and cedar and white fir. The ground cover in such a virgin forest is particularly fascinating and beautiful; along the watercourses and in moister places, fringes of delicate maidenhair fern, and great clumps of the tropical looking sword fern. The smaller plants refined by the deep shade are specially interesting. There is the Clintonia, or queen's cup, with its single turquoise blue berry, one of the purest blues in nature; wild ginger, with its broad green leaves and aromatic root; gold thread, whose roots stripped of bark are of a golden yellow; the pale saprophytic dutchman's pipe, and masses and stringers of club moss. The fallen half-rotten logs are often covered with trailing mats of the lovely little Linnea or twin flower.

"I never cruise a virgin timber tract, destined for sale and logging, without a distaste for the inevitable destruction of so much beauty. As a forester it is my job to grow timber for use of the nation," (Note—Koch does not refer to providing timber for settlers or taking care of local demands first) "when the trees are mature or overmature. But even with the best of forestry practices calculated to reproduce the stand for a new crop, logging the white pine type must inevitably be violent and destructive. Log chutes are built up the lovely small courses, eroding out their channels; the bulldozers relentlessly plough through, opening up logging roads, tearing great gashes in the hillsides; the giant tractors wade without mercy through underbrush, fern and saplings, dragging mighty loads of big logs, till the ground is ripped and torn to a welter of dust or mud. It is not a pretty sight."

Concerning logging camps, Koch said "I have spent many days in logging camps, necessary in the course of my work, but I never did like them. Logging is at best a ruthless and violent business. The camps are temporary and purely utilitarian, and always a scar in the midst of the forest, leaving apparent destruction behind them."

Koch was obviously torn between his professional training and a desire to preserve some of the conditions that existed at the turn of the century when he was scouting lands and adding Forest Reserves. It should be noted that his following statement was written before the Wilderness Act was proposed. How well did Koch know Robert Marshall, a founder of the National Wilderness System? The following indicates they had similar thoughts about preserving some areas for the future.

Koch wrote: "I sometimes go over our old timber sale operations, and while as a forester I may be highly gratified at the success in re-establishing a fine stand of seedling pines, offspring of the seed trees left, still I cannot help recalling what the virgin stand looked like, and knowing that it will be a hundred years or more before we again have a big forest, and of course we will never again grow three or four hundred year old trees like the ones we have cut down. If we hope to retain any such areas for our future pleasure, they must be specially protected as parks or natural areas. We have a few virgin timber areas set aside in the white pine forests of Idaho, and notable progress has been made in withholding remnants of the noble redwoods in California and the matchless coast Douglas fir forest of the Olympic Peninsula. As timber grows scarcer, pressure will inevitably come from the industry to log these remnants. There is much misleading talk about selective logging and the possibility of having our cake and eating it too, but the fact remains that by no forestry measures can a virgin forest of big timber be logged without destroying much of its natural charm."

Harvesting old-growth timber creates changes that will never be restored in a managed forest.

Koch was particularly hard on members of the timber industry. To him, "The lumberjack is the natural product of the hard and tough methods of a logging

Opposite: Two views of a loading tractor skidded white pine logs with a McGifford Loader. This was in 1928 in the Bobtail Creek area, Kootenai National Forest near Libby, Montana.

In the early days of the U.S. Forest Service it was lumberjacks horses and steam engines that provided the muscle to get logs to the sawmill. Here, in 1926, logs are being skidded by horses to a landing on State of Montana lands in the Kootenai National Forest. The logs were then loaded on railroad cars by the donkey engine.

228539

Empty cars being pushed into position for loading on a narrow gauge railroad in the Kootenai National Forest in 1926.

A steam engine and A-frame loading logs in the Kootenai National Forest in 1926.

Loading logs on cars in 1926.

Log loader at work in 1926 loading railroad cars with logs from skidways at the bottom end of a chute. This was in the Calahan Creek Sale Area, Kootenai National Forest.

A railroad engine also used by Polly's Lumber Company.

Logs cut on lands owned by the State of Montana being loaded by a J. Neils Lumber Co. steam engine near Libby, Montana, in 1926.

By the 1920s, caterpillar tractors ("cats") and logging trucks were rapidly replacing horses and trains, and internal combustion engines were replacing steam engines for all uses. Here in 1926 a caterpillar tractor hauls logs to the end of a logging railroad.

A Caterpillar 60 delivering logs to a landing in 1930 near Pierce, Idaho, and the Clearwater National Forest.

operation. I have many good friends among lumbermen, but years of work in the big timber do something to a man. The atmosphere in a logging camp is not a cheerful one. The lumberjack is rarely a light-hearted individual, but inclined to be silent if not morose, and the foremen take on pretty much the same characteristics as the men. One hears little of laughter or jokes in a logging camp. The men eat their enormous meals silently, and sit in the bunkhouse or outside them, tired from the heavy work that necessarily goes with the job of man-handling big logs. It is not that the men are worn out by over-hard work, but heavy manual labor.

"I often contrast the atmosphere of the logging camp with that of a ranch in the cattle country. Cowboys sing and laugh occasionally, and joke with each other; the burden of life sits easily on them. Possibly it is in part due to the difference between living in a heavy timber country from the open range.

"I notice the difference even in our Forest Service rangers. The men in Idaho or western Montana in the heavy timbered country are inclined to be sober and serious like the lumberjack. Their work throws them in contact with the logging operations, and on top of that they have the mental and physical strain of a three month's fire season every year, which surely tends to sober a man beyond his years. Perhaps working with horses and being much in the saddle has something to do with it as well. The packers, even in Idaho, are more like the cowpunchers; I always like to sit and visit with a bunch of packers."

Hired as a tree planter in May 1927, Lloyd Donally had to hike 15 miles by trail to his first job location. The salary was 53 and one-eighth cents per hour. However, 40 cents was deducted for each meal. All supplies, equipment and seedlings were brought in by mules. Quarters were army pyramid tents, six to eight men to a tent. Bunks were made by using two six-inch poles of split cedar placed on the ground about 30 inches apart and filled with straw. Blankets and a tarp were furnished—no pillows or hand towels. The work was hard, but the food was delicious and you could have all you wanted.

Living conditions for early logging crews were as rough as the work. This, in turn, shaped the mood of most crew members. They were quiet, surly and sullen.

Donally notes, "In the fall of 1926, an epidemic of spinal meningitis had hit the camp on Big Creek. Two men died. At that time the men slept on the ground with whatever they could find to put under them—boughs, ferns, grass, etc. It was decided this was not good, healthwise, and in 1927 bunks were made as previously described."

By 1930, Donally had a 14-man crew and his salary was $110 per month and board. In the Fall of 1931 he had a 40-man planting crew and his salary was $125 a month including board. According to Donally, "Tree planting is rough, hard work; climbing steep hills over windfalls and through thick brush. Very little level ground is involved. Men were expected to plant 1,200 trees per eight-hour day. (It takes 1,200 trees planted on a spacing of 6 feet by 6 feet to cover one acre.) I believe I sent more men down the road talking to themselves who were unable to meet this quota than on any Forest Service job I was ever on. Anyone who ever worked under my supervision on tree planting never forgot how I routed them over the hills."

In 1934 Donally had a 60-man planting crew. He notes that because of the depression, these men, like those the year before, stayed on the job. "These men were from all walks of life, railroad men, lumber men, farmers, etc., going to work wherever they could get a chance. In 1935 the crews were "CCC boys." The average number of trees planted per man per day dropped to between 400 and 600. According to Donally, ". . . I have never seen boys who could stow away food as they did. At mealtime everything was cleaned up. The cook couldn't seem to put out enough food. One morning at breakfast a boy ate 32 hotcakes and another 28. It apparently did not bother them as they went

A donkey skidder and loader used by the Polly's Lumber Co. To move, the skidding cable was anchored and the machine pulled itself forward. Note the wood cut and ready to feed into the boiler.

A donkey skidder and loader used by the Polly's Lumber Co. To move, the skidding cable was anchored and the machine pulled itself forward. Note the wood cut and ready to feed into the boiler.

A steam engine loading flat-bed Northern Pacific railroad cars.

Using a tractor and an "A-Frame" to load a truck required more agility and skill than raw muscle power.

A tough way to load logs—rolling them up skids onto a truck with cant hooks.

After logs reached larger lakes, they were often towed to sawmills by tugboats, like this one in Coeur d'Alene Lake.

Logs piled along the banks and waiting to be floated downs the Stillwater River in 1924 near Olney, Montana.

Chutes were often used to move logs many miles to mills or transportation points. This log chute was in use in 1923 on the Kaniksu National Forest.

When there was not enough slope, horses were used to "trail" the logs along the chute as here in the Lolo National Forest.

In 1923 Loggers using "cant hooks" break into a skidway in the Lolo National Forest and roll logs into the chute where they are hauled away by a team of horses.

A new log chute in the Lolo National Forest showing how the logs were hewed that make up the chute. Using chutes to move logs also kept dirt from being ground under the bark. Saws in the sawmill would get dull much faster if logs were skidded on the ground.

"Trailing" ponderosa pine logs in 1922 in a chute constructed by the Montana Logging Co. The chain is hooked to the last log in the line. Note the teamster rides on the end of the log.

immediately to work. After being there two weeks these same boys couldn't eat over six. With good food as an incentive, these boys would pitch in and do their part rather then go back to the main CCC camp."

About the end of the CCC program in the Spring of 1942, Donally had two 12-man CCC crews. Donally provides this description: "These crews were fairly good tree planters, but of all the CCC boys I worked with, these possessed a real streak of orneriness. They would steal, lie, were destructive, and out for a fight. Later I learned some of them were former jailbirds from back East."

According to Carl Kreuger he spent the summer of 1927 helping map areas for planting that were burned in 1926, a bad fire year. According to Krueger, "At that time the only three species of trees being produced were western white pine, western yellow pine and spruce. So on the maps we designated western yellow pine for the warm dry sites; spruce for the wetter areas and white pine for the north and east slopes and deeper soiled areas on all exposures. In practice this worked out well except for the yellow pine where there were many poor plantations. The failures were more due to wrong seed source than to poor site selection. The spruce and white pine, especially the white pine, did very well until blister rust came along later."

Krueger's description of tree planting in 1927 supports the account of Donally. "There were between thirty and forty tree planters in camp divided into three crews. We regularly planted a thousand or more trees a day, and got up to 1,200-1,300. We were in a big burn and planted from creek bottom to ridge top. One day a man on the slope above me dislodged a log which rolled down the hill. The planter next to me and I threw ourselves behind another log, and the rolling one went right over our heads.

"The planters were closely checked. A consistently poor planter, or one who was unable to keep up with us was sent down the road pronto."

Planting trees - starting a new forest - was an important job from the very beginning in the National Forests of the Northern Region.

Kreuger also mentions after the 1927 planting season he was sent to the Falls Ranger Station where he helped operate a white pine seed extraction plant. Cones were bought from locals at 75 cents a sack. Kreuger's pay was $60 per month plus board and ". . . we did all our own cooking."

In 1928 Kreuger was a member of a timber stand improvement crew on the Kaniksu National Forest. There job was ". . . to follow the loggers, pile the brush for burning, cut small grand fir and hemlock and girdle the big ones and generally open the stands up for white pine reproduction. It worked, too; many fine stands of white pine resulted on the Kaniksu and elsewhere."

Concerning tree planting, Elers Koch provides the following: "One of the activities which has given me great pleasure and satisfaction as a forester is tree planting. As a forest supervisor of the Lolo Forest I early established the Savenac Nursery, which at one time was the largest of the Forest Service nurseries, and which for 35 years produced from two million to seven million young trees a year. The great 1910 fires left millions of acres of mountain country blackened and devastated, and it had been our job for years to restore such areas to productivity by artificial planting. Perhaps one reason why tree planting is a pleasant job is because the spring planting is one of the earliest field activities of the year. As soon as the snow goes off the burns, in late April or May, the planting crews are put out. When the weather is good, as it often is, May is one of the pleasantest months of the year to be in the hills. The sunshine is warm, there are no flies and mosquitoes, the spring flowers are in bloom, and all the trees and shrubs are dressed in new fresh green.

Scalp a spot, make a hole, put in a tree, heel in the dirt around the roots, climb six feet to the next spot, repeat the process - and do this all day. This is the way trees are planted and new forests started.

"On a big planting operation, tent camps are established, usually for about 50 men, the mules are brought in from winter pasture and the pack trains commence to move in supplies and equipment and bales and bales of burlap-packed bundles of young trees from the nursery. The planters work in crews of 15 men, moving in echelon across the burned area, their mattocks rising and falling as they skillfully set the little trees, only six to eight inches high, in the moist receptive ground.

"If one is fortunate one may come back again in 30 years and walk under the crowns of a splendid pole stand of vigorous young trees. Some of my oldest plantations are nearly 35 years old" (now they are approaching 80 years old), "and it is always a joy to see them and watch their development year by year. When a Forester goes to St. Peter for a final accounting, I am sure that when he is asked what he has done with his life, if he can point to thousands of acres of once ugly burn now covered with a growing forest, it will offset many sins of omission and commission."

Blister Rust: The White Pine Killer

It was a pitiful time when most of the big, magnificent white pine trees in the Northern Region were reduced to grey, dead skeletons by the blister rust fungus - an unwelcome guest from Europe.

The light, strong and straight-grained lumber made from white pine is a favorite wood of carpenters, especially cabinet makers. It is easy to work and it readily takes most types of finish. Since the first settlers came to the shores of North America, white pine has been in high demand. Today there is not enough white pine to meet demands. As for value, in 1989 a board foot (one inch×one inch×one foot) of white pine stumpage sold for as high as $3.50—about the same price you would pay for a thousand board feet of stumpage if you bought it during the first decades of this century.

As settlers in North America moved west, loggers moved parallel with them through the white pine forests in New England and the Lake States. Until they hit the Great Plains, magnificent white pine lumber was available to settlers for construction.

The forests of giant, virgin white pine trees in New England and the Lake States were almost all reduced to boards when the Forest Reserves were created in 1891. The only extensive stands remaining were western white pine in western Montana and northern Idaho, and sugar pine, a white pine relative, found along the Pacific Coast.

The national demand for white pine lumber was still very high when the Forest Reserves became National Forests in 1905. By 1945, however, white pine was rapidly disappearing as a commercially valuable species for timber production. It was a disease, blister rust, not over-harvesting, that reduced the great western white pine forests to insignificance.

Blister rust is a fungus that came from Europe to America in a shipment of nursery stock in 1909. In 1910, in a shipment of European pine nursery stock, it also came to Vancouver, British Columbia. As smallpox and measles raised havoc among the Indians in North America, blister rust raised havoc with white and sugar pine.

The blister rust fungus requires two hosts for survival. One is gooseberries or currants, members of the genus Ribes, which the fungus does not kill. The other is white or sugar pine, which is almost always destroyed.

Spores of the fungus are wind-blown from the Ribes and enter a pine tree through its needles. The infection then travels down the branches to the trunk of the tree. Here cankers form that eventually girdle the tree and it dies. Spores from the cankers on the white pine are wind-blown and infect the Ribes but cannot infect other white pine. The cycle must include the Ribes.

By 1923, blister rust was infecting trees in north Idaho. Within the next few years, many forest stands were peppered with skeleton-like gray ghosts that had once been beautiful and valuable white pine trees. Surely something could be done to stop this destruction!

Since Ribes are essential to the life cycle of blister rust, removing them breaks the cycle and eliminates the disease. In 1924 an experimental blister rust camp, the first of many, was established in the Upper Priest River drainage in Idaho. Crews of young men were hired to remove Ribes plants. Before long, blister rust camps containing from 30 to 200 persons were established every summer throughout the range of the western white pine. By 1945, almost all stands of trees that were composed of 50 percent or more white pine trees, millions of acres of them, had been worked over one or more times by blister rust crews. Between 1933 and 1941 as many as 12,000 persons were employed every summer on blister rust crews in the National Forests. After the Civilian Conservation Corps program began in 1933, many blister rust crews were made up of CCC enrollees.

Individual crews, usually consisting of about 10 members and a checker would form a line and march about six to eight feet apart through the white pine stands looking for Ribes. They would frequently march a mile or more through very rugged terrain before making a return trip. The person on the far end of the line held a large ball of light cotton store-string on a stick. This was unraveled as the crew moved forward. It served as a guide for the return trip to be certain the area was thoroughly examined. For several years after an area was worked over by a blister rust crew, a principal material in many birds nests was store-string from the guide lines.

When one of the crew discovered a Ribes plant, he would shout "bush." The crew would stop while the offensive plant was removed and hung in nearby branches to dry. Metal hooks were used to remove the larger plants, smaller ones were pulled by hand. It was the checkers job to follow the crew to be certain all the Ribes and their roots were removed.

During World War II, the young men who would have made up the blister rust crews were mostly committed to fighting the war. So German war prisoners, Mexican nationals, high school students and delinquent youths were used, but there were not enough of them to accomplish much. Significant progress in the effort to fight blister rust stopped until the war was over.

And the Ribes battled back. Gooseberries and currents are tough survivors. If a root remnant is left in the ground, a new shoot soon springs up from this source. Tiny, new seedlings can escape detection and in a few years form new thriving plants to spread blister rust spores.

The struggle to save the magnificent western white pine is still continuing at the time of this writing and there is another fascinating story to be told about this conflict in the years between 1945 and 1990. But the blister rust camps today are only places to find artifacts. Students from all over the United States are no longer recruited to make up crews who will spend their summer, struggling over steep, mountainous terrain; swatting mosquitoes; pulling spines from devil's club out of their hands; nursing the many stings from irate hornets; being bored by looking intently at the ground for hour after hour; eating meals that contain almost 4,000 calories a day; feeling grimy from fighting fire whenever needed; and saving money so they can afford to return to college. Not long after the end of World War II the cry of "bush" ceased to be heard in the western National Forests.

"Doping the chute" with crude oil in the Meadow Creek area of the Kaniksu National Forest in 1923. This was in the Hedlund Box Co. timber sale. Doping was done to reduce friction and make the logs slide easily. In winter, snow served this same purpose.

This "splash" or "flush" dam in the Coeur d'Alene National Forest used a flume and water to flush logs off the mountain. Logs were piled on the skidway in front of the dam, water was released into the chute and logs pushed in as long as the water lasted. The dam was then closed until there was enough water again to float more logs.

Training new scalers and providing refresher courses is essential if the Forest Service expects to keep the highest possible degree of accuracy. If the scale is too high, the sawmill owners are mad—if it's too low, the taxpayers are mad. If neither one is mad, scalers worry about their accuracy. This is the future for these scalers attending a scaler's school.

Forest Care

Slash that was not removed by loggers or those gathering fuel, was tightly piled and burned. This slash disposal work was in the Skookum Creek area of the Coeur d'Alene National Forest.

An early timber sale in the Bitterroot National Forest with slash (limbs and unmerchantable parts of cut trees) all piled and waiting for a good time to burn.

More slash piled and ready to burn from an early timber sale in the Deerlodge National Forest.

Christmas trees from the National Forests have always been an important crop. Here bundles of Douglas fir Christmas trees are being made in 1939 at the Hofert Company' Yard at Darby, Montana.

Next, the bundles of trees were loaded on trucks for delivery to the railroad station.

Finally the trees are loaded in railroad cars and shipped to markets "in the East."

A tree planter in the Big Creek area of Idaho's Northern Panhandle, 1925.

A 1938 tree planter in the Lolo National Forest "wicking" his trees. This is a term given to the process of separating trees from bundles and laying them on wet burlap so the roots don't dry out.

There were attempts, mostly with unsatisfactory results, to plant tree seeds. Here a crew in 1911 is using corn planters to plant ponderosa pine seeds in the Savenac area of the Lolo National Forest.

Planting crew, Martin Creek, Cabinet National Forest, 1921.

Since many of the planting sites were remote burned areas, mules were needed to transport camps, food and tree seedlings to these areas. Here Bert Henderson has the mules loaded and ready to move out in 1917 at the Powell Ranger Station, Clearwater National Forest.

On a spacing of 6 feet×6 feet, trees are planted in 1925 by a Forest Service planting crew in a burned area of the Big Creek drainage, Idaho Panhandle National Forests. Planters need strong arms, good legs and a supple back—but foremen and inspectors could be a little portly.

The importance of planting trees and restoring valuable forests was a major message in early Forest Service public information releases. Here, in a hands on demonstration, a Boy Scout studies the rocky ground for a place to plant a tree in the Lolo National Forest.

Camps for tree planters were constructed in remote locations. Tents were erected for sleeping, cooking and eating, and storing trees and other supplies. Overhead usually had separate and tents erected some distance away from the tents used by the crew. Above is the Hemlock Ridge Planting Camp on the Clearwater National Forest in 1939. LLOYD PARKER PHOTO

An Indian huckleberry camp in the Kootenai National Forest in August, 1939. This area was maintained by the Forest Service for huckleberry pickers. The annual harvest from this area was 20,000 or more gallons of these delicious berries.

Wood from the National Forests has always been important for fuel to heat homes. Here ranchers are gathering firewood from a timber sale area after the logs have been removed. Removing this material also reduces the danger of serious wildfires.

In 1921 many horses and lumberjacks were needed to keep logs moving at the sawmill. Here logs are being unloaded from an eight-wheel truck at the Neil Lumber Co. yard in Libby, Montana.

Unloading logs into a hot pond near Henderson, Montana. This was an extremely hazardous job and there were many broken bones and even deaths when logs were released and thundered erratically off the railroad car. Note the man who has released the binder chain is ready to jump if he has to.

By 1941 the overall size of the Anaconda mill at Bonner, Montana, was about as big as it would ever be. Over the years, however, many changes have been made to incorporate new technology and the development of new markets. The present mill, owned by the Champion International Corporation, has been recently updated to better process smaller logs. There are few available large logs left in this area.

In the early days no sawmill complex was complete until it included "company houses," like those in the foreground. These houses were usually owned by the company and provided near the mill for key overhead. Bunkhouses were normally provided for common laborers. The strong "Hellgate Winds" that blow at this junction of the Clark Fork and Blackfoot Rivers are a great asset for drying stacked lumber.

In 1940, Findell's Sawmill, Missoula, Montana, belches smoke, steam and sawdust as logs enter one end of the mill and boards come out the other.

In 1910, the Anaconda Mining Company operated this sawmill at Bonner, Montana. It provided all the lumber needed for the Company's mining operations in Butte, Montana. In addition, the millions of board feet of lumber surplus to the Company's needs was sold on the open market.

In 1936 in spite of the depression, the Neils Bros. sawmill at Libby, Montana, was operating full blast.

In 1923 the Headlund Box Co. maintained this camp near Meadow Creek in the Pend Oreille National Forest. Such camps were a basic ingredient for creating logging legends and tall tales.

In 1920, slow, but effective, a water wheel, similar to the first sawmills in the east, turns a sharp, circular saw and makes boards for a mining operation near Elk Summit in the Bitterroot Mountains.

Thirty-five million board feet of white pine logs waiting to be sawed into lumber in 1938 at the Potlach mill in Lewiston, Idaho. Western white pine at this time was the most sought after lumber in the Northern Region of the Forest Service.

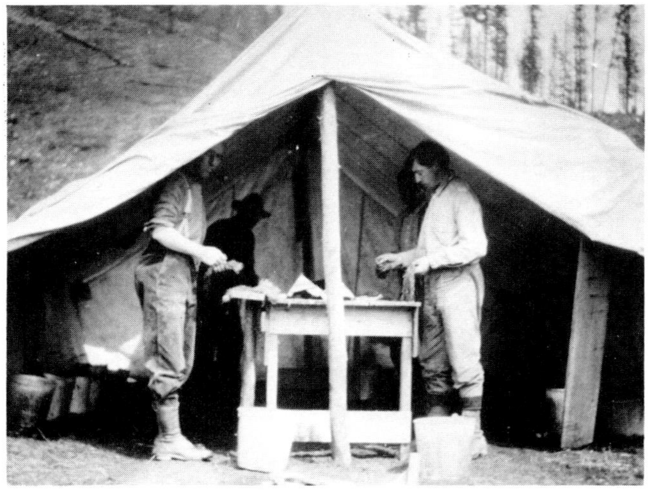

Seedlings being placed in a "transplant board" at the Savenac Nursery in the early days. Preferred planting stock at this time was called "2-2 stock." This meant it was 2 years in a seed bed and 2 years in a transplant bed.

The Savenac Ranger Station and bunkhouse in the winter of 1910—before the Big Burn

In 1910 the Big Burn eliminated the buildings at the Savenac Nursery and tents were erected as temporary quarters. Smoke from this fire is still obscuring the mountains in the background.

In 1912, trenching paths in the transplant beds at the Savenac Nursery was done to insure adequate irrigation water reached the roots of the transplants.

Seedlings being transplanted by crews at the Savenac Nursery in 1912. The nursery is still in its early stages of development.

The Savenac Nursery in 1941. This view includes part of the south portion and main buildings on U.S. Highway 10. In the center background is the older part of the nursery along Savenac Creek. This nursery site was in the old Cabinet National Forest and is now in the Lolo National Forest.

A closer view of the main buildings at the Savenac Nursery in 1941.

The group that initiated the establishment of the Priest River Experimental Station. This picture, taken in 1911, shows a far different area than today's visitor would see.

Soon after the U.S. Forest Service was created, experimental stations were established to learn more about forests and the relationships between plants, animals, soils, insects, diseases and other influences. The buildings above were located at the first Priest River Experimental Station, Priest River, Idaho in about 1913.

An important objective of the Experimental Forests in the Forest Service is to share the knowledge they have gained through their scientific experiments. It is a safe conclusion these men left this 1916 Ranger's Meeting at the Priest River Experiment Station with more wisdom than they had when they arrived.

CHAPTER TWO:
Fire

A burned over area on the Flathead National Forest, 1926.

Fire

For 800,000 years fire has been a major force influencing plants and wildlife of the National Forests in the Northern Region. This includes both the kinds and amounts that can exist in any area. For the past 11,000 years the area of the Northern Region has been in a postglacial period – and wildfire has been common during every warm, dry summer. Every acre in the Northern Region that is not rock, ice or water has periodically burned in the past.

Many plants and animals in the area are dependent on fire for their health and continued existence. For example, lodgepole pine trees and redstem ceanothus reproduce best after a burn. Many plants thrive on nutrients released by fire. These nutrients, in turn, help create healthy deer, elk and other grass or shrub-eating wildlife. Some trees, Douglas fir and larch, develop thick bark that spares them from being killed by frequent ground fires.

The interval between fires depends on location. The interval between fires on dry, south, mountain slopes is about 10 to 30 years. Fuels in these areas will usually get dry enough to burn every year. On the colder, north facing slopes that hold moisture better, the interval between wildfires is about 80 to 200 years. In most years, fires on these slopes are too wet to burn.

When dry, the pitchy evergreen forests and dry grasses in the Northern Region are ideal fuels for generating wildfires. In drought years, lightning storms that have little rainfall start many fires. Before efforts were made to put them out, fires that started in June or July, often burned until snow covered the ground weeks or even months later. Because there are many more lightning strikes in Idaho, there are also more fires in this part of the Region.

To the first Europeans, traders, trappers and mountain men, the occasional fires that swept the land were of little interest. They did not have much effect on game or furbearers. To these rugged people, fire and burned areas were only a nuisance.

Shortly after the Civil War, big changes began in the West. The Indians were moved to Reservations and cattle replaced the buffalo. Minerals, especially gold, silver, and copper were discovered in the Rocky Mountains and booming mining towns erupted. Sawmills were soon making lumber for thriving new communities. For those tough enough to settle here, free land was available.

By 1900, periodic wildfires were no longer just a nuisance. It was becoming necessary to fight the wildfires that every so often blackened rangelands and forests. The grasses and shrubs were needed to sustain horses, cattle and sheep. Grain crops needed protection. Sawlogs were needed for making timbers and lumber needed for developments – for creating a new civilization. Allowing large trees to be destroyed by fire was an unnecessary waste.

Fire was also a serious threat to life itself. The settlers were familiar with some of the disastrous wildfires that had such tragic results in previous years. For example:

–The 1871 Peshtigo fire in Wisconsin burned over a million and a quarter acres and killed some 1,500 people.

–The 1881 fires in Michigan that burned over a million acres and killed 138 people.

–The 1894 Hinckley, Minnesota fire that, by comparison, only burned 160,000 acres. However, it also destroyed 12 towns and killed 413 people.

In 1905 fire was an unwelcome visitor to the brand new National Forests.

Harper's Weekly 1871.

The past toll of lives and resources lost to wildfires was tragic and unacceptable. Something had to be done to stop them. Who should be responsible for preventing them or putting them out? The *National Geographic* magazine issue of September 1898, provides an insight into the public mood at the turn of the century concerning wildfire in the new National Forest Reserves. The article also contains a suggested solution. Authored by Richard U. Good and titled "BITTER ROOT FOREST RESERVE," the article contains the following information about this Reserve which was located in western Montana and Idaho:

> "The most striking feature is the large portion of it (the Bitter Root Reserve) that has burned over, nearly all of it having been visited at different times by fires and at least one-third of the standing timber having been destroyed.
>
> "The foregoing illustration depicts a scene of which all Americans should be ashamed. The aborigines held this region for many ages as a sacred trust transmitted from generation to generation. They recognized its beauty and utility and did naught to impair the grandeur of the one or the permanence of the other. And what has the Anglo-Saxon done? As a community is visited by a devastating scourge, as a face is disfigured by some foul disease, so have the forests been visited and disfigured by him. Reaping where he has not sown and failing to restore where he has destroyed, a noble heritage is slipping away through carelessness and cupidity. A hunter or traveler leaves his campsite fire unextinguishd, a herder starts a fire in the fall that the course grass may be burned and in the spring be replaced by a tender growth which is more nutritious to his flock, the prospector burns the undergrowth that the mineral-bearing rocks may be uncovered, the result is that thousands of acres are devastated."

Concerning "The large areas that have been burned over," the author continues:

> "It is a debt due to posterity that the damage be repaired. This end can be accomplished only by a systematic effort under proper direction. The Yellowstone National Park and the Yosemite Park have for several years past been patrolled by troops of cavalry of the U.S. Army, who have not only been able to keep watch on the class of people to whom these fires are usually traceable, but, by going promptly to localities where smoke is visible, have been able to extinguish with little exertion fires which, if left alone, would have in a short while have devastated large areas. In some European countries it has been found necessary in order to produce certain results in reforestation to transport soil in baskets by the hands of men to form a new covering for the naked rock, so that vegetation may be reestablished. It is not probable that we shall be reduced to such extremities in this country, but we should resist all influences that have a tendency to produce such a condition."

"Wildfire," a word that creates anxiety in everyone's mind, is a familiar term in the National Forests of the Northern Region. The sight of fire in the forests and grasslands, the acrid smell of woodsmoke, tearful irritated eyes and a red sun glowing through a smoky haze are common during hot, dry summer days.

It was stories like this describing the horrible effects of fire and the need for their control—even if it required the use of the military. Accounts like this helped lawmakers decide to move the Forest Reserves out of the U.S. Department of the Interior and placed under Gifford Pinchot and his Professional Foresters in the Department of Agriculture. After the U.S. Forest Service was created in 1905, prevention and control of wildfire in the Forest Reserves and National Forests became a Forest Service responsibility. The public wanted wildfires controlled, and they expected action. This issue could not be taken lightly.

Gifford Pinchot responded to public concern about wildfire in the 1907 publication "The Use of the National Forests:"

TO PROTECT AND GROW MORE WOOD FOR USE

> The national Forests occupy high mountain lands, rough and rocky, and which will always be of value chiefly for the production of timber and wood. The first thing that is made sure is that the timber is not burnt up; the next, that it is used, though not used up. Before there were any National Forests enormous quantities of the people's timber on the public domain every year went up in smoke. Forests which covered districts as large as the State of Rhode Island were completely wiped out in the course of a few days. It meant losses to the people of millions and millions of dollars. Fire destroys quickly; trees grow slowly. It often takes a hundred years to make good the damage done by a single day's fire.
>
> In National Forests there is a force of men on the ground whose business it is to look out for fire. They have been remarkably successful in keeping it down. Since the fire patrol was started less than one-third of one percent of the total area of the Forests has been burned over, and the money loss has been insignificant. This is a wonderful improvement over the old conditions on the public domain, where fires were incessant and enormously destructive.

In another section:

FIRE

> The people have helped the Forest Officers immensely in preventing and fighting fire. There are not half enough rangers to suitably protect the Forests. The only way to keep fire down is for everyone to take a hand at it. Look out for small fires; they start big ones. See that camp fires are completely out before leaving them. Never burn brush or dangerous slashings in dry or windy weather. The time to fight fire is right at the start. When it once has a good headway it is often impossible to control it. As soon as fire is discovered put it out if you can, and in any case notify the nearest Forest officer at once and give him all the assistance possible.
>
> Both State and Federal fire laws apply in National Forests. Persons who start fires intentionally or through carelessness will be most diligently followed up, arrested, and prosecuted to the full extant of the law. The penalties are severe.

Pinchot believed the best way to keep forest fires small was to get to them faster by building better access to National Forest lands. This meant building an extensive system of roads, bridges, trails and telephone lines.

Finally, [in another section] protecting lands and resources from fire is used as a reason for developing the National Forests:

IMPROVEMENT WORK

> Nothing will do more toward giving the National Forests the best kind of protection against fire, and nothing will help more to open up their resources for everybody's use than the construction of a great many well-built roads, bridges and telephone lines. Easy and quick communication to all parts of a Forest must be had if fire is to be kept down. The settlers, prospectors, miners, lumbermen, and stockmen profit directly from all work of this kind and can be of great assistance in pushing it through.

The stage was set, controlling fire was, and still is, the most colorful and demanding aspect of all management efforts undertaken in the National Forests of the Northern Region. Soon after it was created, the brand new U.S. Forest Service had to take a seat in Mother Nature's "School For Wildfire Suppression." Lesson one was "The Big Burn of 1910." It was the kind of lesson that would never be forgotten. After this lesson, preparing for and fighting wildfire was about to become a way of life during the hot, dry Northern Region summers.

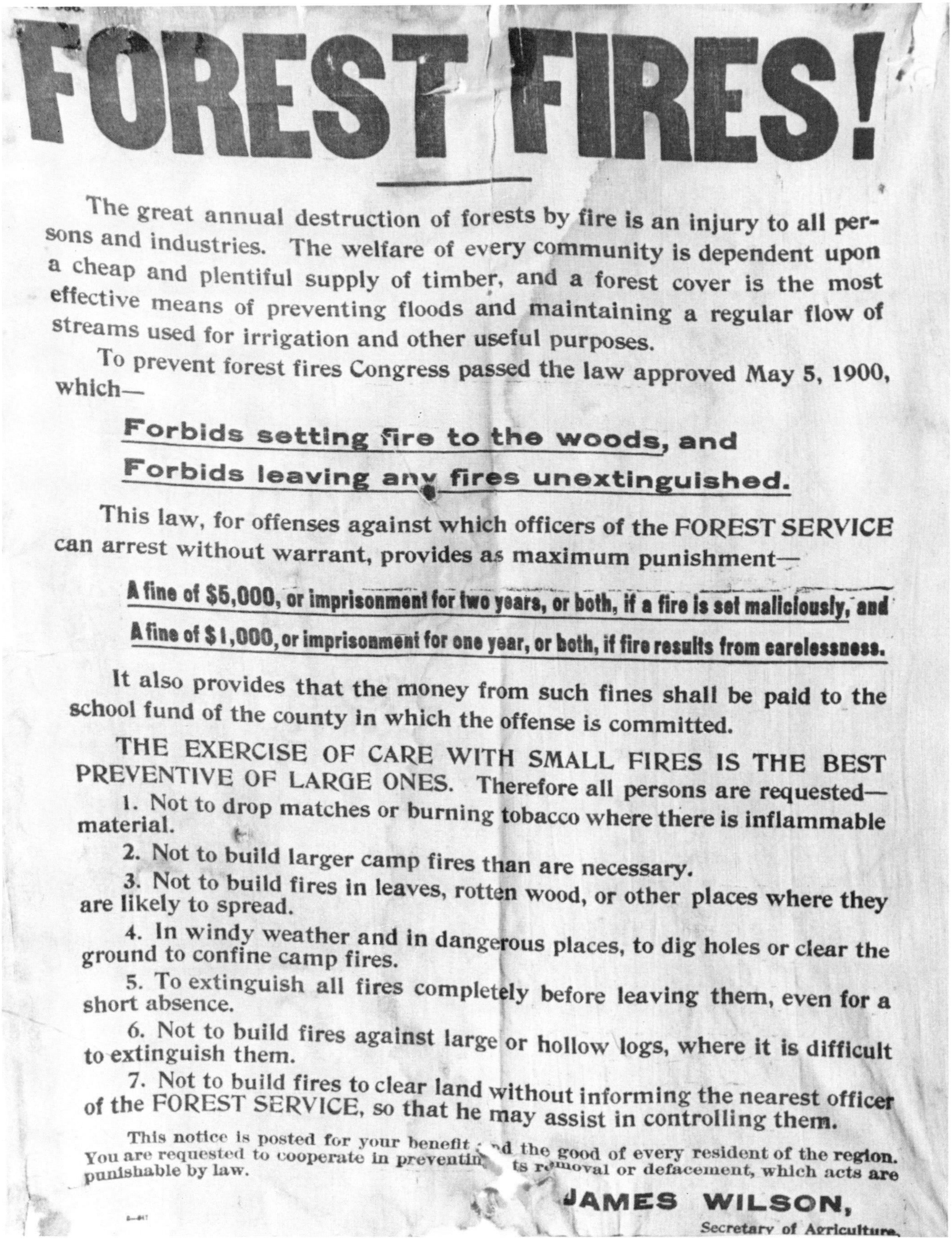

FOREST FIRES!

The great annual destruction of forests by fire is an injury to all persons and industries. The welfare of every community is dependent upon a cheap and plentiful supply of timber, and a forest cover is the most effective means of preventing floods and maintaining a regular flow of streams used for irrigation and other useful purposes.

To prevent forest fires Congress passed the law approved May 5, 1900, which—

Forbids setting fire to the woods, and

Forbids leaving any fires unextinguished.

This law, for offenses against which officers of the FOREST SERVICE can arrest without warrant, provides as maximum punishment—

A fine of $5,000, or imprisonment for two years, or both, if a fire is set maliciously, and

A fine of $1,000, or imprisonment for one year, or both, if fire results from carelessness.

It also provides that the money from such fines shall be paid to the school fund of the county in which the offense is committed.

THE EXERCISE OF CARE WITH SMALL FIRES IS THE BEST PREVENTIVE OF LARGE ONES. Therefore all persons are requested—

1. Not to drop matches or burning tobacco where there is inflammable material.
2. Not to build larger camp fires than are necessary.
3. Not to build fires in leaves, rotten wood, or other places where they are likely to spread.
4. In windy weather and in dangerous places, to dig holes or clear the ground to confine camp fires.
5. To extinguish all fires completely before leaving them, even for a short absence.
6. Not to build fires against large or hollow logs, where it is difficult to extinguish them.
7. Not to build fires to clear land without informing the nearest officer of the FOREST SERVICE, so that he may assist in controlling them.

This notice is posted for your benefit [illegible]d the good of every resident of the region. You are requested to cooperate in preventin[illegible] ts removal or defacement, which acts are punishable by law.

JAMES WILSON,

Secretary of Agriculture.

An early fire poster explains why forest fires are not wanted and cites stiff fines for those who are careless with fire. The notice was directed by Secretary of Agriculture, James Wilson, who had this position when the Forest Service was created in 1905.

The Big Burn of 1910 and Other Fires

One of the most spectacular wildfires of this century, "The Big Burn," stretched from the Salmon River in central Idaho to the Canadian Border. It was indeed a baptism by fire for the new, fledgling U.S. Forest Service. Most of the Big Burn was in Forest Reserve lands.

It was early in 1910, before the Big Burn, the new U.S. Forest Service started to organize the fighting of wildfires in the Northern Region. They were soon to experience one of the worst wildfire years to ever occur in the Northern Rockies.

The 1910 wildfire season began in March when unusually high temperatures were recorded. Record highs continued into April. In June, one of the normally wet months, there was only one-half inch of precipitation. Fires, mostly started by lightning, were soon burning all over north Idaho and western Montana. No rain fell in July. Some of the fires remained out of control.

Roy A. Phillips, who was employed by the Forest Service in 1910 recalls, "Foresters as a class had very little if any training as fire fighters. Trails were few. Discovering fires was the result of patrols, often seen only after fires had burned several days and columns of smoke had drifted high in the sky. Patrolmen tried to make some high point immediately after lightning storms, but this often meant fighting dense brush up a steep mountain for hours at a time. Often when men reached a fire, they were too worn out to fight it and had no food supplies, or effective tools to work with."

The young U.S. Forest Service was baptized with fire in 1910. The "Big Burn of 1910" was the most disastrous fire ever recorded in the Northern Region.

In August, the U.S. Weather Bureau classified the area where the fires were burning as being "in a state of unprecedented drought." Nights, however, were getting unusually cold. While many fires were still burning, there were firelines completed around most of them. There was still a need for more firefighters.

Anyone wanting to fight fires was hired. The pay was 25 cents an hour. Skid rows in nearby cities were almost empty as winos became temporary firefighters. Good crews, however, composed of hard workers were recruited from area logging camps and mines. On August 8, President Taft alerted the Army to stand by and provide help if and when requested by the Forest Service.

William Greeley, District One (Northern Region) Forester for the U.S. Forest Service, told reporters "We are holding our own against all the fires." He was optimistic, after all, it was now late summer. Surely the Fall rains and snow would begin any day.

Instead, on the afternoon of August 20, an unprecedented wind hit. It struck first near Elk City, Idaho. Then, with hurricane force, drove a wall of flames, 30 to 50 miles wide through northern Idaho and western Montana. This invincible firestorm crossed roads, rivers and even mountain ranges without hesitation. It ignored the struggling firefighters and their feeble control efforts.

On August 22, the mighty wind died down. The monstrous fire stopped running. By the end of August, it was all over—the fire was out. It was time to take stock of what happened and lick the wounds. They were deep. About three million acres were incinerated and over seven billion board feet of timber charred—enough to build over 50,000 four room homes. About one third of Wallace, Idaho was lost. Many lives had been saved through a dramatic evacuation by railroad trains of people living or visiting in this area. Other communities that suffered losses included Avery, Grand Forks, Kyle and Falcon, Idaho, and Taft, DeBorgia, Henderson and Haugen, Montana.

Firefighters, Lolo Hot Springs, Montana Area, 1910

ALL PHOTOS BY C.H. SHATTUCK, LOCHSA RANGER DISTRICT

Axemen and sawmen working in advance of the fire.

Firefighters at dinner.

Firefighters following fire line back to camp.

Pack train on the way to the fire at Packer Meadows.

Fire camp at Lolo Pass.

Fire line near the Bullion Mine, Wallace, Idaho.

St. Joe Basin burn area.

Disinterred bodies prepared for transportation and burial on the St. Joe National Forest, Sept. 1910.
JOE HALM PHOTOS

Eighty-seven people gave their lives to this storm straight out of hell. Seventy-eight of these were firefighters and seven were civilians. Seventy-two firefighters, one prospector and two civilians lost there lives in and near Wallace in the Coeur d'Alene National Forest. Four firefighters died in the Cabinet National Forest and two in the Pend Oreille National Forest. Three homesteaders died near Newport, Washington, in the Kaniksu National Forest. There were hundreds of injuries, many very serious. ,

Because of smoke from this fire, daylight was definitely reduced from Saskatoon, Saskatchewan, Canada in the north to Denver, Colorado in the south. Smoke was apparent in Watertown, New York in the east and 500 miles out in the Pacific Ocean to the west.

Joe B. Halm, who had a crew fighting the fire, was overrun by the fire storm and lived to describe it. According to Halm:

> "The wind had risen to hurricane velocity. Fire was now all around us, banners of incandescent flames licked the sky. Showers of large flaming brands were falling everywhere. The quiet of a few minutes before had become a horrible din. The hissing, roaring flames, the terrific crashing and rending of falling timber was deafening, terrifying. Men rushed back and forth trying to help. One young giant, crazed with fear, broke and ran. I dashed after him. He came back, wild-eyed, crying, hysterical. The fire had closed in; the heat became intolerable."

Halm's crew took refuge in a small mountain stream. Here they threw buckets of water on each other while their clothes steamed and smoked. There was some air they could breathe just above the surface of the water. Trees crashed around them. Halm continued:

> "After what seemed hours, the screaming, hissing and snapping of millions of doomed trees, and the showers of sparks and burning brands grew less. The fire gradually subsided. Words were spoken. The drenched, begrimed men became more hopeful. Some even sought tobacco in their water-soaked clothing. Another hour and we began to feel the chill of the night. The hideous, red glare of the inferno still lighted everything; trees still fell by the thousands. Wearily, the men began to drag the water-soaked blankets from the creek and dry them; some scraped places beneath the fallen trees where they might crawl with their weary, tortured bodies out of reach of the falling snags. The wind subsided." It was a long night for the weary firefighters.

Ranger Joe Halm's crew at Timber Creek, headwaters of the St. Joe River, before the great fire in August, 1910.
OREGON HISTORICAL SOCIETY

Haun's Ranger Station in the Lolo National Forest near Haugen, Montana, in the path of the great fire.

Then, Halm described the area when light came:

> "Dawn broke almost clear of smoke, the first in weeks. Men began to crawl from their burrows and look about. Such a scene! The green standing forest of yesterday was gone; in its place a charred and smoking mass of melancholy wreckage. The virgin trees, as far as the eye could see, were broken or down, devoid of a single sprig of green. Miles of trees, sturdy forest giants—were laid prone. Only the smaller trees stood stripped and broken. Inwardly I gave thanks for being alive. A big fellow, a Swede, the one who had refused to stay, slapped me on the back, 'You save me my life,' he said simply. His lip trembled as he walked away."

Seven firefighters were burned alive when they crowded into the homesteader's root cellar at the bottom of the photograph, the night of Aug. 20, 1910, near Big Creek on the St. Joe National Forest.
JOE HALM PHOTO

In addition to Halm, Rangers Phillips, Watson, Vandyke, Rock, Bell and many others saved hundreds of lives by their cool timely judgment. Of the firefighters that died, most perished as they tried to outrun the merciless flames. Some were killed when they were struck by falling trees. Some sought refuge in a shallow hole in the earth. They perished when the cool oxygen in this cavity disappeared and was replaced by superheated air. William W. Morris, another crew leader fighting fire during the Big Burn, stated:

> "When the wind blows hard and other conditions are right, the fire leaps up into the crowns of the trees, and travels along at a great pace, by brands blowing ahead of it and starting new fires. When it is quiet, one may work quite close to the fire, but let the wind blow, and it is necessary to keep a respectful distance."

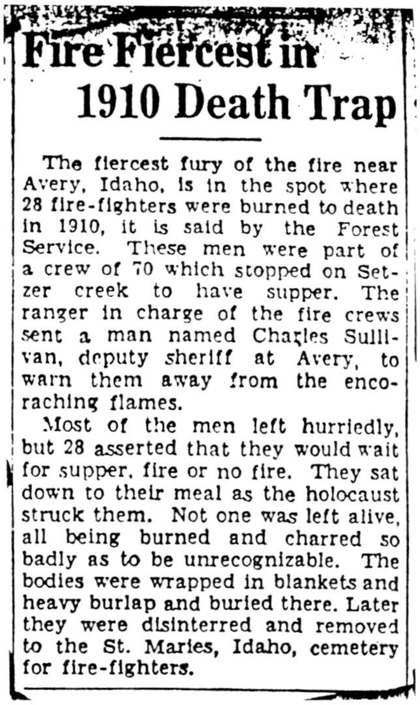

Fire Fiercest in 1910 Death Trap

The fiercest fury of the fire near Avery, Idaho, is in the spot where 28 fire-fighters were burned to death in 1910, it is said by the Forest Service. These men were part of a crew of 70 which stopped on Setzer creek to have supper. The ranger in charge of the fire crews sent a man named Charles Sullivan, deputy sheriff at Avery, to warn them away from the encoraching flames.

Most of the men left hurriedly, but 28 asserted that they would wait for supper, fire or no fire. They sat down to their meal as the holocaust struck them. Not one was left alive, all being burned and charred so badly as to be unrecognizable. The bodies were wrapped in blankets and heavy burlap and buried there. Later they were disinterred and removed to the St. Maries, Idaho, cemetery for fire-fighters.

The high winds of August 20 also struck Morris and his crew as they renewed their efforts to contain a fire. Soon, however, they were surrounded. With all their efforts being in vain, they fled through the fire to safety. Luckily they all escaped and reached the railroad tracks near a small schoolhouse, the Graham Creek School. Morris writes:

> "As we lay down on the tracks exhausted, I remember hearing the school bell give a final clang as it fell to the ground, its supports being eaten away by the tongues of fire. As the fire came, timber was uprooted at just about the time the flame struck it. The falling trees caused a continual roar, and in some places were thrown in rough whirls, resembling the work of tornadoes."

The following stories were in the Missoula, Montana, *Missoulian*, on Sunday, Aug. 21, 1910:

FRIGHTENED PEOPLE FLOCK FROM TAFT

> Telephone messages from Taft at 2:00 AM this morning stated that the forest fires had crept down upon the town and that the Northern Pacific was taking the residents of the famous camp down to Saltese. Two trainloads had left by 2:00 o'clock, at which time the fire had become so serious that it was thought that further wire communication would be impossible. The telephone operator who sent the message said that he had heard refugees from Wallace were also being taken to Saltese. The fire was said to be burning close to their tracks, and it was feared that traffic might possibly be seriously delayed.

Taft was one of the mountain railroad towns destroyed by the fire. The hotel was the only major building that survived. During the construction of the Taft Tunnel in 1908 the town had a large population but by 1910 it had declined to several hundred people.

Another story said:

FIRE-FIGHTERS DIE IN FLAMES

> Kalispell, Aug. 20.—(Special)—Two camps of fire-fighters are reported burned this morning near Kalispell. Word was received from the camp this morning to the effect that 30 men and a number of horses are unaccounted for. Supervisor Haines expressed hope that the men were safe, but stated tonight that the two camps, including bedding and provisions, were entirely consumed. Provisions were ordered rushed from Whitefish to a point near the camp.

Five recovered bodies were those of men on Ranger Pulaski's 40-man crew. This crew was fighting a fire near Wallace, Idaho, when the wind struck. Their only chance for survival was to take refuge in a shallow mine tunnel. The tunnel entrance was at the bottom of a steep canyon. Uprooted and blown over trees slid down the slope in front of the entrance and burned as though they were in a giant fireplace. Soon almost all the oxygen from the mine shaft had been drawn into this inferno. When breathing became hard, the crew began to panic and fought to get out. If they left, however,they would surely die.

Ranger Edward Pulaski near the entrance of the War Eagle Tunnel where he and crew were trapped during the fire.

Pulaski held them at gunpoint and made them lie down with their faces next to the damp tunnel floor. This was the last source of oxygen. Pulaski remained near the tunnel mouth in the most exposed position. It wasn't long before all was quiet.

After several hours, the burning fuel was about gone. Pulaski regained consciousness just in time to feel one of the men crawl over him and say "Too damned bad, the Ranger is dead." All but five were soon breathing normally. Efforts made to save those five were not successful. Pulaski nearly lost his sight from exposure and glare because of his efforts to guard the mouth of the tunnel. When they returned to Wallace, Idaho, the day after their brush with death, they were a welcome sight. They had been given up as another lost group of firefighters.

The nation was shocked by the lives lost and resources destroyed in the Big Burn. The adolescent U.S. Forest Service had indeed been taught a lesson in what wildfire could do. They were determined to stop something like this from ever happening again! Before long the Forest Service philosophy for fighting fires became "HIT 'EM HARD AND KEEP 'EM SMALL." This philosophy, however, was not always possible. It depended upon Mother Nature's providing enough precipitation to avoid drought. But she did not always cooperate. In addition to 1910, there were other lessons to be learned in the bad fire seasons of 1919, 1926, 1929 and 1934.

According to Elers Koch:

> "The year of a forester in the the Northern Region is divided into two parts, the forest fire season and the rest of the year. In a bad fire season from the first of July till the September rains, one lives, breathes and thinks nothing but fire, and a good part of the balance of the year is spent in making plans and preparations for the next fire season."

Koch felt The Big Burn of 1910 was ". . . a complete defeat for the newly organized Forest Service forces." In explanation for this remark, Koch writes "Not only was there the immediate loss, but it is a well known fact that fire breeds fire, and for many years the 1910 burns, with their inflammable dead timber, were a constant threat." He did not think this threat diminished until there was a good new growth of trees on the area.

There would be few disagreements with the timelessness of Koch's assessment of a fire organization:

"Fire fighting is perhaps the nearest thing there is to war, and not without its dangers, and always requiring extreme physical exertion, long hours, lack of sleep and constant nervous tension. Moving in to a fire which has gone beyond control, with a crew of men, when one looks up toward his destination and sees an enormous threatening thunderhead crowding up into the sky, indicating a wild crown fire, it is understandable that one sometimes has a feeling of butterflies in his stomach. In a way it is like troops moving in the front line, and listening to the artillery and shell bursts ahead."

The War Eagle Tunnel entrance on Placer Creek near Wallace, Idaho, where five men died.

Trail crewmen cutting a trail through the burned timber in the Little North Fork of the St. Joe River drainage in the St. Joe National Forest, Sept. 1910.

Missoula photographer R.H. McKay (left) and Ranger Joe Halm (right) posed for a photo in the awesome destruction in the Little North Fork of the St. Joe River area on the St. Joe National Forest. McKay was one of western Montana's most prolific photographers for many years.

The "loop" on the Milwaukee Railroad on the North Fork of the St. Joe River. Most of the this basin and the trestle were burned by the fire. The railroad suffered considerable damage to tracks, bridges, and buildings located along the Idaho-Montana divide routes. The coal-burning engines caused numerous fires along their right-of-ways, so the railroad electrified its line across the mountains in 1913.

Downed timber near Falcon, Idaho, looking southeast, September 1910. The destruction of the heavy stands of timber in the forests of Idaho was incredible and complete. It would take decades to bring this forest back to production again. Some effects of the fire are still noticeable today.

Burned timber on Rainy Creek, a tributory of the St. Regis River near the east portal of the Milwaukee Railroad's tunnel.

Koch provides a good description of a fire that could not be held. All possible efforts were made to insure the safety of the crew in this, one of the most dangerous jobs in the world:

A lookout platform, located on the Bitterroot Divide near the Taft Tunnel along the Idaho-Montana border, is about to topple from the effects of the fire. It is a wonder that the telephone poles are still standing.

> "I recall one particularly fierce battle in the Hemlock Creek fire in the Clearwater National Forest. The fire, which was largely in dead timber, had blown up almost immediately after the lighting struck. Trail crews and other available men were moved in from all directions. I happened to be on the Forest, and seeing the burst 30 miles away, headed for the fire as fast as my horse could take me. I went in with a fire camp on the lower point of the fire.
>
> "The first day a line was partially completed around the lower edge of the fire, but it was unfortunately necessary to locate the line on a steep slope with the fire backing down to it from above, which is always a hard line to hold.
>
> "The dead timber stand was heavy, with giant white pines 150 feet high, and much of the half rotten hemlock and white fir to throw sparks. In midafternoon the humidity dropped and the wind blew, and the fire raged. Our puny 15 foot wide line below it seemed a very slim defense. We stopped constructing any new line, and all hands concentrated at the danger points to hold what we had. Dry logs on the fire side of the line were blazing wildly, and every man with a shovel was throwing dirt desperately to try to damp down the fierceness of the blaze. Burning snags were falling constantly and our position was one of the greatest danger. My heart was in my mouth every time I heard a snag crack.
>
> "One tremendous dead white pine four feet through, and perhaps 160 or 170 feet tall, was all ablaze, and everyone was watching it. It started to topple. 'Look out, here it comes!' and the crew scattered both ways along the fireline. Down it came with a thunderous crash that shook the earth, and slid down the mountain, top first, in a cloud of dust and smoke, across our fire line, and stopped a hundred feet below. Of course it carried fire of all sorts with it. I got the scared crew together, and working desperately with saws, axes and mattocks, we looped a line around it. Scarcely had we completed this line when another snag came plunging across the line a hundred feet away. Again I rallied the worn out crew, and again we looped a line around the break. But it was no use; a third and fourth snag came down. The line was clearly untenable, fires were burning wildly on both sides, and it was clear that there was a general blow-up. The foreman and I got the crew together, checked to see that we had the last man, and retreated down the mountain toward the camp."

Joe Bouchamp's homestead on the Middle Fork of the St. Joe River, where ten firefighters burned to death on Aug. 20, 1910.

Grand Forks, Idaho—a rough-and-tumble town between Taft and Avery. It was to be in the direct path of the flames on August 20.
OREGON HISTORICAL SOCIETY

According to Koch, accidents were not always prevented:

> "Falling snags are the greatest danger in working along the fire line. In fires in which I have taken part, at least six or eight men have been killed by burning snags crashing down on them, and it is not a pretty sight to see a man with his head smashed and the brains spilling out. It is extremely demoralizing to a crew, and after such an accident one always expects a big walk out of men the next morning."

Koch did not have much respect for early, untrained firefighter crews.

> ". . . in the earlier days of the Forest Service we had to gather together such men as we could from the streets, saloons and off the freight trains. As a result, there was almost as much misery connected with handling the men as with fighting the forest fire. Such transients were almost like children, unreasonable, irresponsible, and acting purely on impulse. They had absolutely no feeling of obligation or responsibility to their employers, which was perhaps natural since few employers at that time felt any responsibility toward them.
>
> "If we were lucky there would be enough good lumberjacks in a crew to fill the axe and saw gangs, and the punks and stew bums would be given a shovel or mattock to get the best they could of it.
>
> "I remember an old ranger calling me on the phone and saying a fire had broken over badly and he needed some men right away. 'How many do you need?' I asked. 'Send me 10 men; that is if they wear hats; if they wear caps, I'll need 30.' The distinction was that the respectable lumberjack never wears a cap in summer, always a felt hat, usually a black one, whereas the pool hall boys usually wore caps, and shoved their hands deep in their pockets."

Koch also provides an interesting story about pickup crews:

> "One night in 1910 we were shipping out all the men we could scrape up from Missoula for fires to the west. I had 50 to get off at Frenchtown and Walt Derrick was taking another 50 to Avery. The men piled on the train, and being mostly drunk, promptly went to sleep. When we got to Frenchtown, I could not tell which were my men and which were Walt's, so I shoved off about 50 of the men who were awake, and so got the best of Walt, for he got the worst drunks!"

In 1910, the lack of communications for directing the firefighting effort was a major problem. The heliograph was tried but was not good enough. After The Big Burn, and as fast as appropriations permitted, telephone lines were constructed to overcome this obstacle. It required thousands of miles of wire, fastened to insulators and mounted on trees and poles, that provided good communications. These soon linked most Forest Service offices and work centers. The system required a lot of maintenance, but it was worth it.

View of De Borgia, on the main line of the Northern Pacific Railroad approximately 85 miles west of Missoula, Montana.

Pre-fire scenes in De Borgia, Mont., soon to be engulfed by the raging flames.

Destruction Along the Northern Pacific Railroad in Western Montana

Top and middle: Fire refugees in western Montana. Hundreds of people were forced to flee the raging fire in the Coeur d'Alene area, many were brought by train to Missoula, Montana.

De Borgia after the fire of August 20. The saloon on the far left was one of the few buildings to remain intact. The flames destroyed the railroad towns of Taft and Haugan and threatened Saltese, St. Regis and Superior.

THE SEATTLE POST-INTELLIGENCER

VOL. LVIII, NO. 99. SEATTLE, WASHINGTON, MONDAY, AUGUST 22, 1910.—FOURTEEN PAGES.—IN TWO PARTS PRICE FIVE CENTS

FOREST FLAMES WREAK HAVOC; $1,000,000 IS LOSS AT WALLACE AND 25 DEAD

Birdseye View of Wallace, Idaho, Before the Fire

IDAHO AND MONTANA TOWNS SUFFER DISASTER

Great Army of Terror Stricken Settlers Are Driven From Their Homes

WALLACE, Idaho, Aug. 21.—Daylight this morning showed that the imminent danger of the city's destruction has passed, but it brought also confirmation of losses which were only rumored and suspected during the conflagration of the night. At least two died in the city fire, John Boyd, a pioneer of the Coeur d'Alenes and a former Oregon Railroad & Navigation Company agent, and an unknown man or woman, having been incinerated in the Michigan hotel. Only the skull of the latter was found in the ruins this morning.

Toll of Dead Unavailable

Of the fire fighting forces an accurate toll of the dead and wounded is quite unavailable, but the known of the dead number twenty-four, the total injured twenty-five, in addition to ten blinded. The steady work of the city fire department, members of the Twenty-fifth infantry, colored volunteers and the forestry forces alone saved Wallace from total destruction. The conflagration in the east end was stayed shortly after 11 o'clock and back firing on adjoining hills to the west and south prevented [illegible] fires in those directions. It is estimated that the loss in the city is about $1,000,000. The entire eastern section from Seventh street to Canyon [illegible] destroyed, with three terraces of residences on the hillside.

MISSOULA, Mont., Aug. 21.—Mercilessly and relentlessly, the forest fires in Western Montana and Idaho are sweeping over a vast area, driving hundreds of fugitives before them, destroying small settlements and wiping out of existence millions of dollars' worth of property.

The situation tonight is more serious than it was in the early morning, except as to Wallace, Idaho, where it is believed that nearly half the city will be saved.

Communication Broken

Communication with Wallace to the west has been possible at intervals today, but eastward it is entirely cut off, and it is known that the entire east half of the town, above Seventh street, has been burned. [illegible] of that a hard fight is being made, and with an improvement in the water supply there is more chance that the flames may be driven back. [illegible] this afternoon the Daily Missoulian's reporter at Wallace [illegible] a wire. He summarized the situation as follows:

Lives Lost

[illegible]

Elsewhere in the fire zone the situation has gone from bad to worse today. The most serious incident is reported late this afternoon from [illegible] country, where 180 men engaged in the forestry service are missing and it is feared they have been burned to death.

Rescuers Organize

When the fire approached the camp, where there were 200 men, [illegible] of the fighters took a horse and, riding the animal to death, reached [illegible] camp and organized a rescue party, which penetrated the fire to [illegible].

[illegible] of the men were found in the water, where they had gone for safety, and they were unharmed. Of the remaining 180 no word has been received. The forestry service has organized a relief train, [illegible] equipped with pack animals carrying provisions and hospital supplies, [illegible] will endeavor to get through the fire.

Refugees Suffer

About [illegible] refugees have been brought into Missoula today. There is much distress among them. Their wants are being supplied by Missoula people and they have been given temporary homes. The first of [illegible] trains came in over the Northern Pacific's Coeur d'Alene branch, bringing the patients who had been in the Sisters' hospital at Wallace and as many refugees as could find a place in the small train. There were [illegible] on [illegible] train, and a second train at noon brought as many more.

Roused From Sleep

These people came from the small towns along the line between [illegible] and Wallace. Many of them had been roused from their sleep by the [illegible] on the train, whose summons had been the first intimation that the fire was near; there had been no sign of it when the people went to bed [illegible] urday night. In most instances they escaped only with scanty clothing.

Child Born in Car

A woman who had fled from her home at midnight gave birth [illegible]

MANY BELIEVED TO BE CUT OFF BY FLAMES

Twenty Forest Fire Fighters Are Reported to Have Lost Their Lives

SHOOTS WOMAN DOWN IN STREET FULL OF PEOPLE

Thrice Wounded, She Runs and Passersby Seek Safety Behind Poles and Trees

KILLS BROTHER AND NIECE IN SUDDEN FIGHT

Italian Runs Amuck With Knife—Is Shot After Stabbing Many Persons

CAPT. E. B. WOOD IS SLAIN AT SEA BY INSANE PASSENGER

Murderer Then Jumps Overboard From Deck of S. S. Buckman

Wallace, Idaho, was in the direct path of the flames. Destruction was complete in the east end of town—one hotel, the railroad station, an iron foundry, two breweries, numerous businesses and over 150 homes were lost. The fire jumped over the rest of the town, leaving it untouched.

A view of the main part of Wallace showing the courthouse in the center.

Looking east into the main business district.

Near the ruins of the foundry.

The St. Joe Ranger Station near Falcon, Idaho, before the fire, Henry Kottkey was the ranger.
OREGON HISTORICAL SOCIETY

The new St. Joe Ranger Station built on the ruins of the old station, September 1910.

The St. Joe Ranger Station two weeks after the fire. It was totally destroyed.

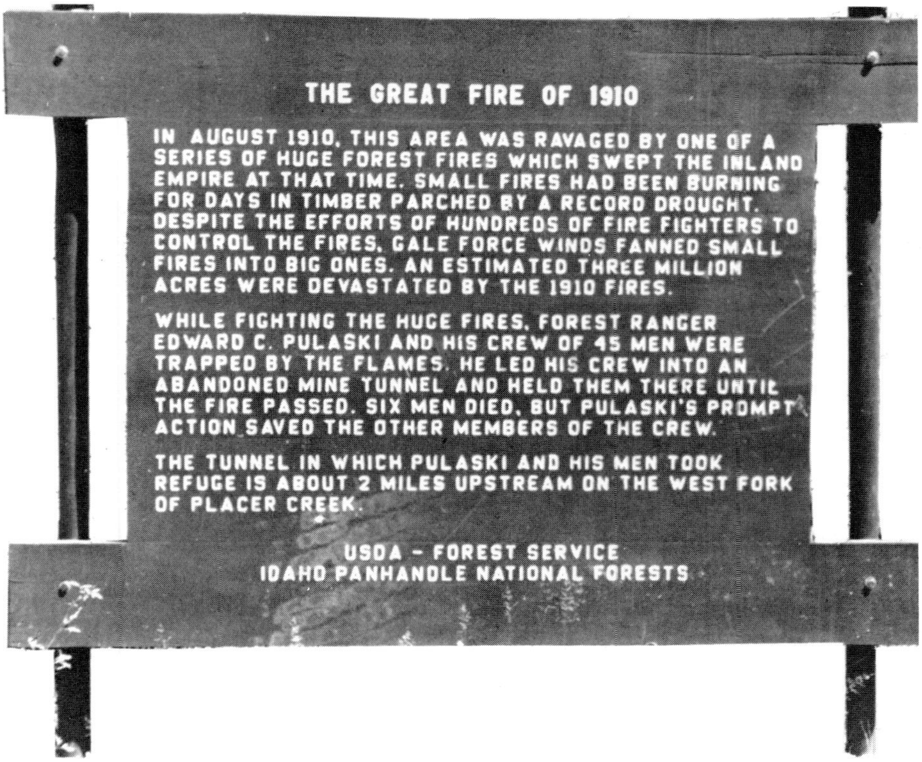

A monument near Wallace, Idaho, was dedicated in 1984 by the U.S. Forest Service to the memory of the dead firefighters.

Grave monument for some of the firefighters located near Wallace, Idaho.

The graves of 54 firefighters at St. Maries, Idaho. A total of 75, perhaps 78, lost their lives in the holocaust of August 20-21, 1910.

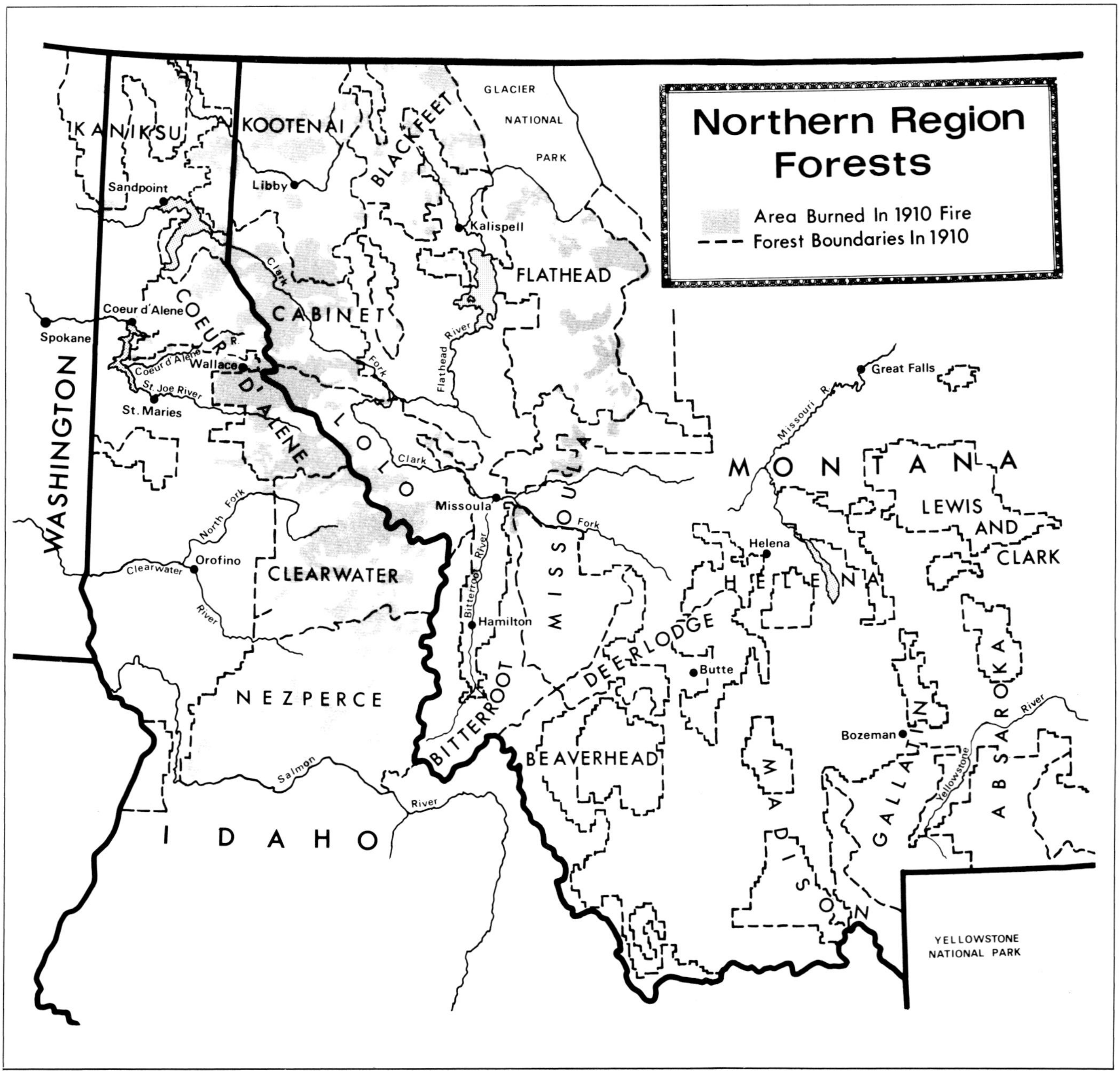

Fire lookouts, at first very crude, were strategically located and added to the telephone network. Observers, "Lookouts," often camped near these points during the entire fire season. In later years, especially during the days of the CCC program, permanent installations with towers and cabins were constructed. Building materials were brought in by pack trains. Some of these lookouts are still in use today. But for those assigned early day lookout points, the Forest Service furnished a tent, a few cooking utensils and a bed, a double bitted axe and firefighting utensils. Often two men were stationed at a point—one lookout and one smokechaser. They built their own fireplace for cooking. Starting wages for a first year lookout were $70 per month, including room and board—and their room was "all the outdoors." Fish, huckleberries and grouse were good supplements to the diet.

It took a special breed of men to fill these jobs and live in such primitive conditions. They loved the mountains and solitude. They possessed the same spirit as the fur trappers and prospectors who had made footprints in these remote areas before them. It was their love of their work and their attitudes that became the backbone for the Forest Service in the Northern Region.

According to Robert G. Elliott, in 1925 a "Smokechasers Rations" consisted of a can of corned beef hash, a can of bacon, a bar of chocolate, two packages of hardtack and some dried rice. However, to supplement this "poor provender," they would mix up "bannock" flour—white flour with some baking powder and salt mixed in. About four pounds of this mixture was carried dry in a cloth sack. About four pounds of dry beans were carried in another sack. They also carried a "big hunk of hard smoked bacon." When they arrived at a fire, they put some beans in the kettle from their Forest Service cooking kit, covered them with water from their canteen, added a few hunks of bacon for flavor, put the kettle on hot coals and "let her cook."

Hours later, when the fire was under reasonable control, they fried some bacon slices. Next they mixed a bit of bannock flour with water into a bannock and fried it in the bacon pan. They were then ready to enjoy a wonderful hot meal of beans, bacon and bannock bread.

In the early days of firefighting, equipping firefighters and getting them to the fires was a tough job. In 1910 the Forest Service purchased all the axes, shovels, saws, mattocks and sougans locally available. Then, these tools, camp supplies and food had to be moved by pack strings to the fire camps—sometimes 70 or more miles from the nearest road and not much improved trail. The mess outfit was fit into two boilers which made a side pack for a mule. Improvements were made as fast as possible. By the early 1930s, the first crude radios made their appearance. Now it was possible to dispatch a crew working in the field to a fire. Provided, of course, they were not in a "dead spot" where the radio wouldn't work, or the batteries hadn't gotten too weak. Soon aircraft were added to the list of available firefighting tools.

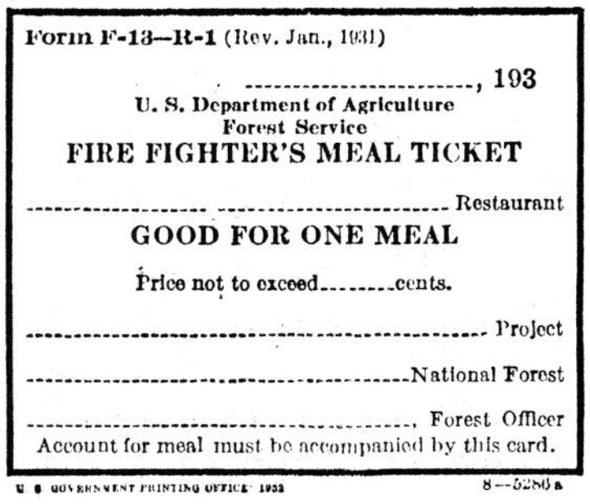

Form F-13—R-1 (Rev. Jan., 1931)

------------------------, 193

U. S. Department of Agriculture
Forest Service
FIRE FIGHTER'S MEAL TICKET

------------------------ Restaurant

GOOD FOR ONE MEAL

Price not to exceed--------cents.

------------------------ Project

------------------------ National Forest

------------------------ Forest Officer

Account for meal must be accompanied by this card.

Keeping firefighting bills and accounts straight was a bookkeeper's nightmare. Filling in the blanks of this "Fire Fighters Meal Ticket" helped. The ticket was for use in the decade of the 1930s. Note the price of the meal was not expected to reach as high as $1.00.

An airplane and crew used in firefighting efforts at the Newport, Washington landing field in 1925.

Ed Thenon, who had been a ranger on the Clearwater Forest in 1906, compared early firefighting efforts with how it was done in the 1940s. He said:

A pack train returning to Avery, Idaho, for supplies for firefighters in 1920.

> "Sure it was different, we not only didn't have no smoke jumpers and no airplanes, but no roads and damn few trails. We didn't have no remount station to send us an organized pack string, and no central warehouse to ship us an outfit ready to go on a fire. Time we got an outfit together, and hired a pack string, and picked up a crew, and walked them in maybe 75 miles to a fire, it was sometimes a week or 10 days before we got a crew on a fire. And the men we had to fight fire with, boy oh boy, it was more trouble fighting the men in those days than to fight the fire. They was always drunk when we hired them, and time and again they used to walk out on us just when we needed them the most, and you never could tell what they would do when the fire got going bad and the flames going up through the spruce trees with a whoosh. There's a heap of difference in men. Some of them is real men that can use their heads, but there's a lot of them driftin' 'round the country that's cattle. That's all they are. A mouth and a belly, but no brains. Just cattle."

It was in the mid 1920s that the potential use of aircraft started to look promising. Leon Hurtt provides an account of early use of aircraft:

> "My first trip by plane came at the close of the fire season in 1924 or '25. Fire Chief Howard Flint sent a De Haviland plane to Grangeville to take the Nezperce and Selway staffs up to test us on locating fires from a plane. I made the first hop without incident to Red River and back in about an hour's flying time, a trip that usually took two days by truck. I noted the speed indicator up to 125 miles per hour as we came in to land! Brandborg's hop circled out over the Seven Devils to frighten and scatter widely a bunch of Hibb's fat steers being driven to market from their isolated Snake River range. They had never heard a gasoline engine. On Clyde Blake's trip, the pilot treated him to a surprise loop-the-loop over Grangeville. Clyde was a little green around the gills when he landed. Jim Urquhart took the last trip of the day and narrowly escaped disaster because the pilot, flying down the Selway River, under low clouds with a small gas supply, failed to take the left turn toward the Camas Prairie where the Selway and the Lochsa joined. After he had followed up the rugged Lochsa canyon for several miles, Jim made him understand that he must turn back and follow the Middle Fork to reach Grangeville. We lined up car lights to illuminate the landing place in a stubble field and were almost as much relieved to see the pair down safely as were Jim and Priestly, the pilot." (Priestly was killed a few days after this episode in an air show in Spokane.) Since then planes have been indispensable to the Forest Service.

Delivering fire rations in 1940 near Eagle Point Lookout, Canyon Ranger District, Clearwater National Forest.

In 1929 Evan W. Kelley, the meanest firefighter in the nation, came to the Northern Region as Regional Forester. He was "a Forester's Forester." Because of his management style and his World War I rank, he was usually called "Major" Kelley. Small in stature but big in action, he gained a "can do" reputation in World War I. A cavalry man, he had commanded the "Forest Engineers" at Bellieu Wood. His main objective as Regional Forester, the top man in the Northern Region, was to "whip" the fire problem.

It was soon after the Major's arrival the Forest Service initiated its famous "10 A.M. Policy." This required managers to commit whatever resources were necessary. It meant that in spite of drought, or wind, all fires, (started in their jurisdiction) had to be under control by 10 a.m. the next day. Kelley's fast and hard personal interest in fighting wildfires resulted in a new firefighter's slogan in the Northern Region, "leave a black line behind you."

During the Major's first fire season in the Region, District Ranger L.M. Stewart tells of getting a request to call Major Kelley after he returned from walking around an 1,800 acre fire. According to Stewart, the Major said, "Ranger Stewart,

Smoke chasers ready to start for a fire in about 1917 in an Idaho National Forest.

Firefighting crew with pack-horses on its way to the Bean Creek Fire in Idaho. Oregon Historical Society

Wildfires have always been a component of the environment in the West. This was a fire that went through a slash area in the Lewis and Clark Forest Reserve in 1904. Note the high stumps. This was usually the result of loggers cutting trees in deep winter snow while crews were wearing snowshoes.

Fire camp for the Squaw Creek fire, Gallatin National Forest in 1937. This was before hardhats and flame-resistant clothing. Note the stove in the background and the double boiler which probably contained coffee. Some claim this boiler was always put on when the camp was started and never removed from the fire until the camp was closed. Water and more coffee were added when the level got low. Some oldtimers say that fire camp coffee and paint remover were about the same thing at the end of the fire.

Fire Camp at Beaner Ridge in 1917.

A crew of firefighters on its way to the Bald Mountain Fire in 1929.

how many men do you need to control that fire by 10:00 a.m. tomorrow?" While Stewart thought this was impractical, the Major's attitude about the 10:00 a.m. policy got through loud and clear. According to Stewart, Major Kelley also created a change in the attitudes of the fire fighting inspectors. Where they used to concentrate on suspected "over-manning" of fires, they now began to concentrate on fires that were not controlled at 10:00 a.m. the next day—and why this had happened!

Kelley's interest in fires was further demonstrated by his occasionally taking over the fire desk or suddenly appearing in a fire camp or on a fire line under construction. Everyone knew the Major believed that when a wildfire blew up, whoever was in charge, Ranger, Supervisor or Fire Boss, had failed. He had a very simple solution—the man responsible was fired. All fire control officers now knew they must adhere to the 10:00 a.m. policy. They could watch their Forest Service career swirl around and disappear down the drain if they could not prove failing to gain control by 10:00 a.m. was beyond the capability of any human being.

Major Kelley, the Northern Regional Forester, had no sense of humor when it came to implementing the 10:00 a.m. fire control policy. This policy required all fires to be under control by 10:00 a.m. the day after they were detected - and there were few reasons Kelley would accept if this did not happen!

Bert Cramer tells the following story that supports this attitude of the Major:

> "In 1931 I had about 300 men camped in a big meadow while we were fighting the Deer Creek Fire. One night, the fire seemed to be coming in on us from all sides, which was unusual. I fed the entire crew before dark and put all men back on the fireline again because it seemed that we were going to have to fight to save our camp. About midnight a tall white pine snag struck one of my men squarely on the head, bursting his head in two pieces and literally driving him endwise into the ground. I pulled the entire crew into camp and told them to hit the hay. I found a bedroll and just passed out. The next thing I knew Regional Forester Evan Kelley was shaking me and peering into my face. It was daylight. As soon as he got me fully awake he said, 'What the hell is wrong—kill one man and order 500 to replace him.' The Major then informed me that the sight of this dead man when they carried him into camp the night before had apparently caused my timekeeper to blow his top, resulting in his jumping into a car and driving to Bonners Ferry (Idaho) and ordering 500 men and equipment for my part of the fire. The Regional Forester was in the office when this order came in and he thought he had better come out and see what was going on. As it turned out we did not need or get the 500 men."

Moose Creek fire camp in the present-day Bob Marshall Wilderness Area, 1919.
Courtesy Mike Westrum, photo by Stecker

A fire spreading in young trees, North Fork of Calahan Creek in the Kootenai National Forest.

Trail through burned timber above Camp Martin in the Clearwater National Forest in 1925.

A burned over drainage.

Results of a 1929 wildfire along the North Fork Road near Bailey Lake in the Flathead National Forest.

A picture taken in 1924 showing an area burned over during the Big Burn of 1910 in the Coeur d'Alene National Forest.

Since early telephone lines were strung on trees, for many miles, it was logical to also have "tree phone booths" such as this one in the Clearwater National Forest in 1925.

The 1934 Pete King Fire looking from Round Top toward Idaho Point, Lochsa Ranger District, Clearwater National Forest.

Being responsive to the 10:00 a.m. policy was not easy during these years. Imagine yourself a Forest Service Fire Control Officer on a Northern Region National Forest when the Major was the Regional Forester. Your position is responsible for getting crews and supplies to all fires. Some, started by lightning, are in very remote areas. A remote fire is spotted by a lookout and you send a couple of smokechasers to put it out while it is still small. The lookout calls again and says the fire is mushrooming, the smoke cloud is growing very fast. Now, based on this information, you are pretty sure the smokechasers didn't get there on time or the fire was too much and got away.

It's time to send reinforcements. More often than not, the firefighter crew you dispatch will have to walk 40 or 50 miles. You also have to dispatch a packstring to bring in tools, supplies and equipment. You must be sure a camp and field kitchen is in place and operating. The crew is going to be tired when they arrive, and will not be able to do much until they have been rested and fed. Then, you hope you have sent the right number of firefighters and equipment to control the fire. And you always have to keep in mind, Major Kelley expects you to have that fire controlled by 10:00 a.m. the day after it is reported! You do the very best you can to see this is done. If the fires is not controlled, you can rely on the Major's query.

Wouldn't it be wonderful if the dispatchers could know exactly how big the fire was and how fast it was spreading? If someone could fly in an airplane over the fire and provide this information, dispatchers would be able to send the right number of firefighters and equipment needed. And, wouldn't it be nice if the fire camp, tools and other materials could be dropped in by parachute instead of instead of being hauled many miles on the backs of mules? And, in the early days, dispatchers, perhaps letting their imaginations run wild, could even visualize firefighters dropping by parachute to fires. Yes, aircraft had some interesting potential.

The first year that airplanes were used extensively to scout fires was 1929, a bad fire year. In this fire season, Northern Region Fire Chief, Howard Flint, piloted by Nick Mamer of Spokane, flew over the Sullivan Creek Fire on the South Fork of the Flathead River in the Flathead National Forest and greatly aided the firefighting effort. His view of the fire was so different from what could be determined from the ground. He could see, for example, the Spotted Bear Ranger Station would probably be destroyed unless something was done soon to protect it. He scribbled a message on a piece of paper, tied a weight to it and had Mamer fly low over the Station. He dropped a message that said in part, "Save Spotted Bear Ranger Station at all costs." Charlie Hash, the Flathead Fire Control Officer, was at Spotted Bear and received the message. He immediately pulled in 500 firefighters, and began wetting down the buildings with Pacific Marine pumps. The Station was saved. It didn't take very many examples like this to prove the value of aircraft to scout fires.

It was the disastrous 1929 fire season that led to the creation of the famous Ninemile Remount Station. During this fire season, the Regional headquarters in Missoula, Montana was dispatching rented pack strings to help the firefighting effort. Before the summer was over, however, all the trained pack stock was committed and they were forced to use untrained animals. A major cause of the lack of horses and mules was the mechanization of agriculture. By 1929 tractors and trucks had about completely replaced the need for these animals. This, along with the supply of packsaddles running out, seriously hindered the firefighing effort—and this was something that Major Kelley, the new Regional Forester, could not tolerate.

During the Great Depression, while there was an abundance of CCC and

A burned area, still smoldering. The blackened dead trees still standing and the forest floor littered with trees that have gone down. The acid smoke from smoldering wood puts lungs into distress and creates a most uncomfortable heartburn. It is hard to believe that abundant life is here waiting only for the fire to die and precipitation to help it spring into new life.

A stand of timber is temporarily converted by fire into a prairie in the Kaniksu National Forest. This picture was taken in the vicinity of Gleason Mountain in 1945. Note the young trees growing vigorously among the skeletons of their parents.

other relief labor, locating and building good lookout towers was a high priority of Major Kelley. He personally initiated a drive to locate and construct all the primary lookouts in the Region. Kelley also decided that living quarters as well as observatories would be on top of the tower. This permitted the lookouts to be on their posts 24 hours a day. Material for the lookouts was brought in by pack trains with prefabbed lookouts being constructed in Spokane, Washington.

The Pinkham Creekers were as tough as the area they lived in. Hiring on as firefighters helped them scratch out a living - even if they had to start the fires! They were a challenge to many Forest Service officers.

While most of the wildfires were set by lightning, trains, brush burners or smokers, there were a few set by settlers who wanted to obtain work as firefighters. This was alleged to be the case now known as the Pinkham Creek Rebellion in the Kootenai National Forest in 1924. For several years a number of fires had been set in the Pinkham Creek area. The settlers here had come from the hills of Kentucky and West Virginia. For them, poverty was a way of life. Some made a living by stealing timber and hacking out railroad ties; some made moonshine; and a few tried raising cattle. Fighting fire, however, gave them a chance to make a few extra dollars and a better than average income for a Pinkham Creeker. Sometimes it was just fun to set a few fires "just for the hell of it—to bother the guvmint men."

However, after 32 fires were set in July 1924, the Forest Supervisor decided enough was enough and closed all Federal lands in the Pinkham Creek area. In addition, seven camps were placed around the valley with armed patrolmen enforcing the closure. This got the attention of the fire-setters for several years.

But the Pinkham Creekers were not all beaten. In 1932, Charley Powell, District Ranger at Rexford, overheard a conversation between two Pinkham Creekers indicating they knew about some planned incendiarism. C.S. Webb, the Forest Supervisor hired an "undercover" man to get in with the Pinkham Creekers. This man finally gained their confidence and was informed August 22 was their date for setting a long string of fires.

The day before the fire setting date, the Forest Service had two or three men assigned in the vicinity of each area to be set on fire. A number of them saw and were seen by the Pinkham Creekers, who realized the Forest Service somehow got wind of their plans. Fortunately, they did not figure out an undercover agent was in their group. Apparently it was enough, however, to put a wet blanket over making plans to light more fires. This episode appeared to be the end of Pinkham Creek incendiary fires.

By 1936 the delivery of fire supplies by parachute was becoming perfected. Use of airplanes for scouting fires and dropping supplies was soon to become standard procedure. This replaced the need for pack strings. Landing strips for aircraft were constructed in some of the more remote parts of the Northern Region.

Left: A horse-drawn plow unit makes fire line on the McLendon Butte Fire in 1934, Nezperce National Forest.

Right: Unloading a horse-drawn plow unit from a truck at the 1934 Pack River Fire in the Kaniksu National Forest.

Bert Morris provided some interesting stories of early smokejumping:

> "The year 1939 was about the first year that anything was tried in the field of smokejumping. Part of this nucleus was stationed at Big Prairie. Dick Lynch is one of the individuals who sticks in my memory. Dick had spent the day haying, without a shirt and had acquired a helluva sunburn. The next day a WO delegation (from the Chief's Office in Washington, D.C.) came in and wanted a smokejumping demonstration. The crew obliged but Dick grabbed a chute and forgot to fully adjust the harness. This is before the day of static lines. He made a free fall for a ways, then pulled the rip-cord. That harness snapped against him when the chute popped. You could see every mark of the harness above his waist that night.
>
> "Carl Nussbacker told me that during those rip-cord days he missed the rip-cord and pulled all the buttons off his canvas jumper."

Ernie Richards provides an interesting account of an air cargo drop in 1940. According to Richards:

> "Supervisor Karl Klehm (Kootenai National Forest) instructed me to go to the Libby airfield and meet a plane that was coming to drop a 100-man fire camp by parachute on the Bull Lake Fire. When I got to the airport, I discovered the pilot was Dick Johnson and the plane was one of their Ford Tri-motors. When Dick saw me he said, 'Ernie, you are just the man I am looking for; I need a cargo dropper. My regular man is sick and I don't have anyone to throw out the cargo at the drop area. The plane was gassed up and when it was loaded, we took off for Bull Lake. Dick circled the drop spot before he had me dropping the cargo. Right after I kicked out the first bundle, I placed one hand against the side of the doorway and looked out to see if the 25-man ration box I had just kicked out was going to land in the camp spot. While I was looking out, the plane hit a down draft and dropped several feet about straight down, and the top of the doorway hit me in the back of the head right at the base of my skull. This slightly stunned me and almost caused me to fall out the door. I managed to push myself back away from the door and then sat down while we circled around for another pass, and by the time we got about halfway around I was all right. I never tried that again because I was lucky I didn't fall out. We got the rest of the stuff out in good shape. It took us two days to get all the material for the 100-man camp dropped in place."

Ed MacKay using a Bosworth Fire Trencher in 1939. Machines of this type had limited application because of tree roots and rocky soils.

In 1941, the first smokejumper training base in the nation was located at the old Ninemile CCC Camp that was no longer in use. The Nation's young men were needed for service in World War II. The people working at the Remount Depot were aware aircraft for transporting firefighters and equipment would soon replace their pack strings. The age for using firefighting aircraft was well underway by 1945, the end of the period covered by this book. After World War II, rapid advancements in the use of aircraft continued, but these need to be covered in another volume.

The Forest Service also learned the beneficial uses of fire. Burning logging slash was one of these uses that had several benefits. It reduced the chances of a wildfire getting away in the dead, down and dry debris; it made an impossible area easy to plant; it released plant nutrients that otherwise would be locked for years in slowly decomposing slash.

Berle Davis provides some interesting information about the 1940 "Troy Slash Burning Project." "It was all logged, all in one piece and they wished to fall all cull and non-salable timber and to burn off these and the logging slash for a fresh start on that land. We got a lot of experience at falling trees and dominoed them. You cut one almost to the falling point and aimed the second tree at it to knock it down and aimed a third tree at the second, etc., until about the fifth one you cut clear down, with luck you knock them all down like dominoes." It didn't always work, and there were dangerous moments as they got "hang-ups" down. When they were finished, "you could walk clear across the area on the downed trees." This area was then burned.

According to Davis, after it was lit, "In about 30 minutes, our smoke gave the A-Bomb mushroom cloud serious competition for size and color, and noise too, as there was a terrific roar as it went on that hot and windless day before the fire. Tornado drafts in the fire area actually rotated end for end trees that were fifty feet long and were lifted twenty feet off the ground in the burn." Some of the people setting the fire had a close call. Davis added, "It was considered a successful burn by the overhead and nobody was lost in it!"

By 1945 National Forest Supervisors needing a firefighter camp, just called one of the Forest Service warehouses located in Missoula, Montana or Spokane, Washington. Then, by express or in trucks, the camp, in 25-person units, complete and ready to load on pack strings or in aircraft, was immediately dispatched. If pack stock was needed, a request was made to the Forest Service Ninemile Remount Depot near Missoula. In 15 minutes pack strings, each consisting of nine mules, a saddle horse, all their equipment and a packer were underway in specially designed trucks. There were trained firefighters and smokechasers available as well as organized crews from the blister rust camps. The CCC firefighters had disappeared with World War II, but there were now organized State organizations and firefighting cooperatives as well as the logging crews. In 1945 most Forest Service employees in the Northern Region were made of tempered steel that was poured from a common crucible heated by intense wildfires.

Using airplanes in firefighting was very dangerous, especially in the mountains of the Northern Region. This is what was left of an airplane that crashed in Roaring Lion Canyon in 1939 while delivering supplies to firefighters in the Bitterroot National Forest.

A special 1936 F-1 fire truck ready for action.

HELP
PREVENT
FIRES
BEAR IN MIND

FOREST FIRES
AID THE ENEMY
Crush out your cigarette
U.S. DEPT. AGRICULTURE, FOREST SERVICE
STATE FOREST SERVICE

YOURS IN TRUST
WE MUST
PROTECT IT
FROM FIRE
HOW TO PREVENT FOREST FIRES

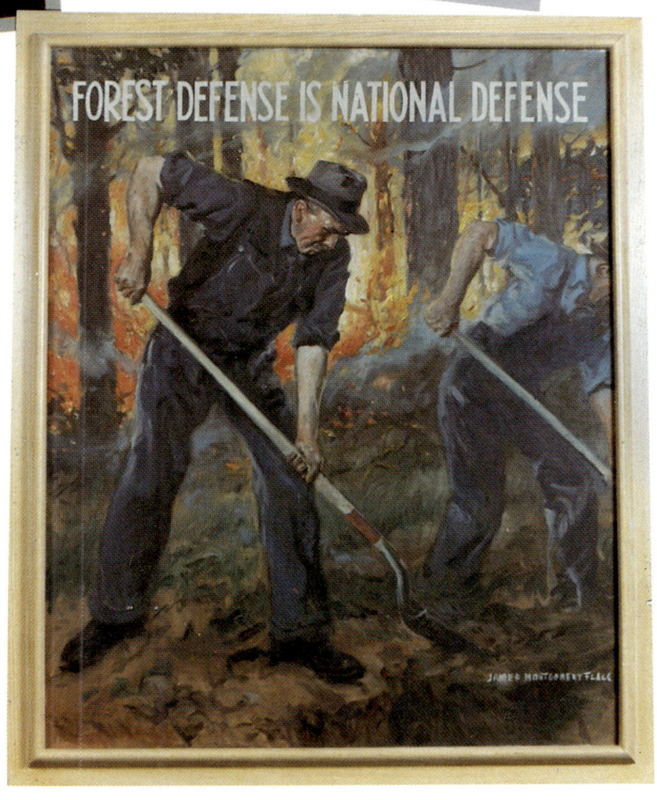
FOREST DEFENSE IS NATIONAL DEFENSE
JAMES MONTGOMERY FLAGG

Finding enough firefighters during World War II was anticipated to be a serious problem. Appeals for civilians to demonstrate their patriotism, as indicated by this poster, were distributed in areas near lands subject to wildfires.

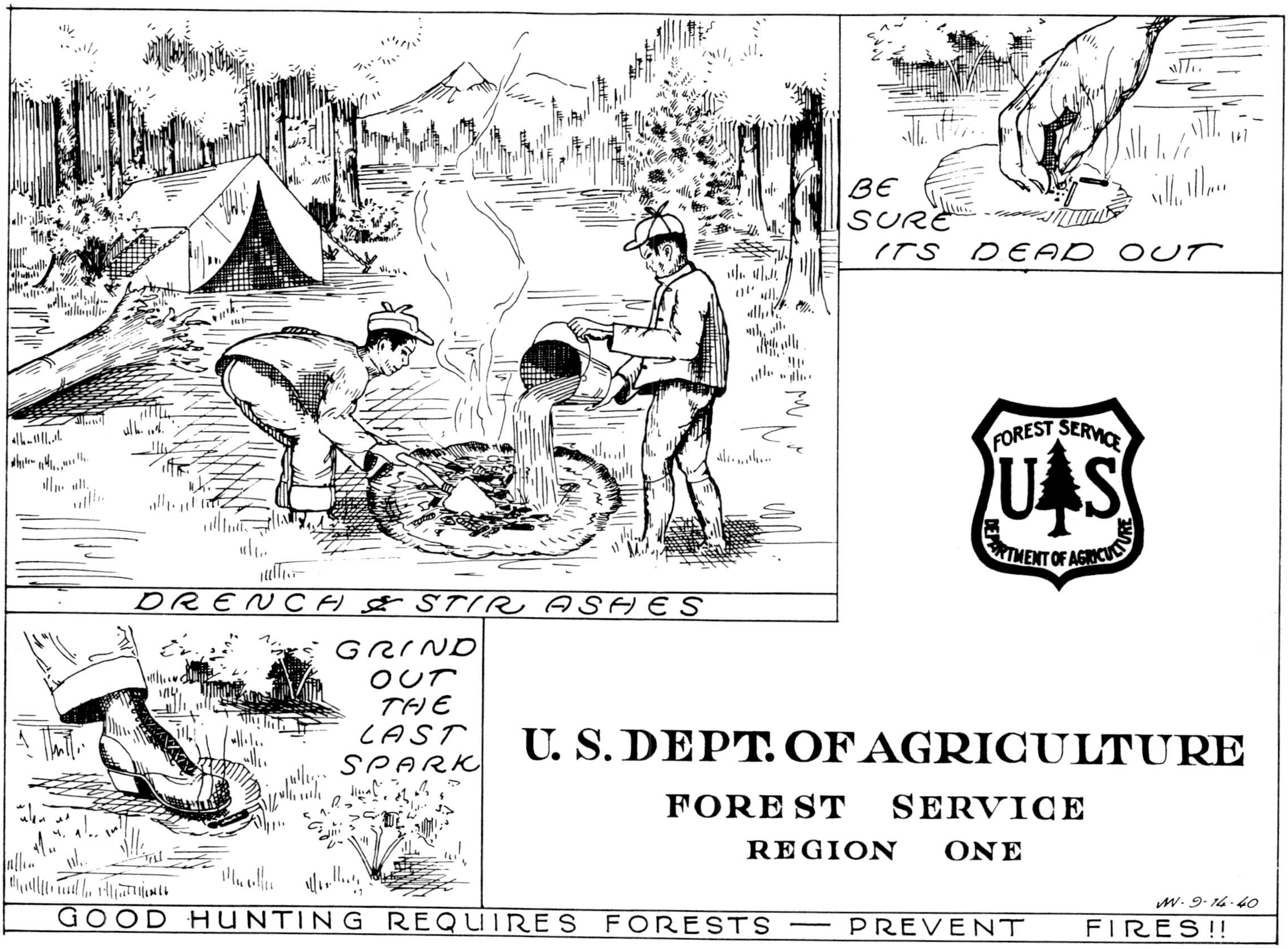
BE SURE ITS DEAD OUT
DRENCH & STIR ASHES
FOREST SERVICE
U S
DEPARTMENT OF AGRICULTURE
GRIND OUT THE LAST SPARK
U. S. DEPT. OF AGRICULTURE
FOREST SERVICE
REGION ONE
GOOD HUNTING REQUIRES FORESTS — PREVENT FIRES!!

GOOD WOODSMEN ALWAYS
LEAVE A CLEAN CAMP
ENJOY YOUR NATIONAL FORESTS
BE SURE IT'S DEAD OUT
DON'T BE A Flipper
SOMETHING TO SHOOT AT
GRIND OUT THE LAST SPARK
"No" HUNTER CAUSED FIRES
U. S. DEPARTMENT OF AGRICULTURE
Forest Service
SPORTSMEN PLAY FAIR
ALL CAMPFIRES OUT
FOR GOOD HUNTING HELP PREVENT FIRES

SPF portable radio set in operation on the Bull Lake Fire, Kootenai National Forest in August, 1939.

Prevent Forest Fires—Your Sport Depends on It

Northern Region fire unit in 1936 at the Nine Mile Camp. *Top (left to right)*: Lewis Smith, Joe Adams, Lawrence Dawson, Jack Demers, Cecil Good, Don Orr, Newt Kirkberg, Bill Pamdell, Gene Meinet, Keith Goombs. *Bottom (left to right)*: McLaughlin, Gordon Stickley, Chuck Nelson, Conecky, Wright, Ervie Northway, Mark Weisel, Bill Kaymeyer, Lloyd Jameson.
LLOYD PARKER PHOTO

Smokejumpers

The feasibility of dropping firefighters by parachute to combat forest fires had been under consideration by Forest Service fire control officials for a number of years. With the rapid improvement of aircraft in the 1920s and early 1930s, the dropping of supplies and equipment proved to be a practical means of supplying firefighters.

The actual idea of parachuting men to fight fire is attributed to a forest ranger in Utah, T.V. Pearson, who experimented with the idea in 1934. After a few demonstrations by a professional jumper, J.B. Bruce, the idea was abandoned as being too risky. The few men who pioneered parachute jumping were considered to be publicity crazy or just plain crazy. The Russian and German armies had perfected mass paratroop drops in the late 1930s, but the U.S. Army had considered paratroopers impractical.

Using parachutes to drop supplies and the potential for dropping men onto a fire could not be ignored, however, and a man appeared on the scene who would be the guiding light for the future smokejumping program. David Godwin was the assistant chief of fire control in the Washington office of the Forest Service and head of the Aerial Fire Control Experimental Project. This was a program begun in California to study the feasibility of dropping chemicals or water from the air to retard forest fire movement. This proved to be impractical with the airplanes in use at the time, but during the years of 1935-39 much was learned about the use of airplanes and cargo dropping in mountainous, timbered country.

Frank and Chet Derry, brothers who were involved in the early development and expansion of the smokejumper program. In 1942 they perfected the "Derry Slotted Parachute" which greatly aided a jumper in steering and guiding his parachute.

Then, serious smokejumping experiments were conducted from October 5 to November 15, 1939, at the Forest Service's Intercity Airport located between Twisp and Winthrop, Washington in the Pacific Northwest Region of the Forest Service (Chelan National Forest). After more than 50 jumps, including many in stands of trees, it was concluded jumpers could be dropped in timber from considerable altitudes, and near the site selected, without injury if proper procedures were followed.

After the successful completion of the jump experiment the previous year, 1940 was the first operational year in the history of smokejumping. There was no increase in funds for the program, but the Forest Service decided to establish two jump training centers, one in the Pacific Northwest Region at Winthrop, Washington, in the Chelan National Forest, and one in the Northern Region at the Seeley Lake Ranger Station in the Lolo National Forest. The Northern Region was particularly suited to the program because of its extreme fire hazards and large roadless areas. In the Northern Region seven smokejumpers were hired and took their initial training at Seeley Lake, 50 miles from Missoula, before being moved to the Moose Creek Ranger Station in the Nez Perce National Forest. The pay for the jumpers was $193 and board for two and one-half months of work.

The training was intense, with special emphasis on physical conditioning. Various developments were made in equipment for greater safety and simplicity. Short-wave radios were added to the jumpers' equipment for signaling to the plane. Improvements were also made in the let-down technique from tree landings.

Nineteen-forty would be an important year in that the first actual fire jump was made by Rufus Robinson and Earl Cooley over Martin Creek in the Nez Perce

A letter from Regional Forester Evan Kelley expresses his displeasure with the idea of dropping men from a plane to fight fires. Just four years later it was proven to be a practical means of fighting fires in remote areas.

Mr. Earl W. Loveridge
Forest Service
Washington, D. C.

July 19, 1935

Dear Earl:

Enclosed is a copy of a letter from J. B. Bruce.

The file in this case is in Spokane, but for purposes of identification of this letter, I will remind you that you wrote some time ago about J. B. Bruce's scheme of dropping men from airplanes for fire fighting. Pearson of Region Four was a party to the scheme.

I am willing to take a chance on most any kind of a proposition that promises better action on fires, but I hesitate very much to go into the kind of thing that Bruce proposes. In the first place, the best information I can get from experienced fliers is that all parachute jumpers are more or less crazy - just a little bit unbalanced, otherwise they wouldn't be engaged in such a hazardous undertaking; accordingly, I discount materially the practicability of Bruce's ideas.

As applied to the average situation, in our country at least, we wouldn' have any great need for dropping men from the air; moreover, it is too risky an undertaking to experiment in. I do not want to be responsible for a lot of compensation cases which, in all probability, might develop from such experimentation in the rough mountain country where Bruce proposes to experiment.

We have had some of the experienced fliers familiar with parachute designing look over the diagram submitted by Pearson. They say that Pearson's design has merit over the regulation chutes ordinarily used today. However superior it might be it wouldn't rescue a man from hanging up on a snag or a tree top, or prevent him from falling off of a cliff if, by some mishap, he should land on such a topographical feature in his descent from a plane.

The point of my letter is that I have no hankering to assume the responsibility for men risking their lives in any such undertaking.

Of course, there are a number of other points to consider. To make the thing practicable at all we would have to have a trained bunch of parachute jumpers who also are skilled firefighters. These men would be called upon for service over a period of possibly two months a year.

I think we can use our available money very much more productively than investing in that kind of service, even admitting the practicability of the scheme. If the Washington Office wishes to carry on further experimentation in this line, it is my request that you assign the project to other quarters.

Very sincerely yours,

/s/ Evan W. Kelley
EVAN W. KELLEY.
Regional Forester

National Forest. Then, at the end of the 1940 fire season, it was determined that a crew of 40 jumpers would be needed for the 1941 season.

The year 1941 turned out to be pivotal for the success of the smokejumping program. Forty jumpers had been authorized, but funds were only allocated for three eight-person squads, or a total of 26 firefighters including a project leader and jumping instructor. Nearly all the previous year's jumpers returned.

It was decided to concentrate the entire program in the Northern Region out of Missoula, Montana, because of the Region's great expanse of roadless areas and the fact that the newly contracted Johnson Flying Service was based in Missoula. They could provide all the planes, pilots and service that would be required for an expanded operation. Crews would be dispatched to other Forest Service Regions as needed.

In 1940 a new base was established at the Stoney Creek Civilian Conservation Corps camp near the Ninemile Forest Service Remount Station. After training, one squad remained at Nine Mile, one was sent to the Moose Creek Ranger Station in the Nez Perce National Forest and one to Big Prairie in the Flathead National Forest. They were then shifted as needed.

A serious outbreak of fires in the Pacific Northwest Region in late July was the only real fire threat of 1941. Only nine fires were jumped. Still, the program had more than proven its worth.

Just when the smokejumper program was being accepted as a major firefighting force the war intervened and seriously curtailed its activities. A four-squad unit was formed but only five experimental jumpers were on hand

Prior to establishing the Ninemile base, a temporary camp was set up for the jumpers adjacent to Seeley Lake Ranger Station, Lolo National Forest, in 1940. The ranger station is on the right. EARL COOLEY COLLECTION

Air strip at Seeley Lake, used for some practice jumps by CPS jumpers during the war.

Earl Cooley, left, and Jim Waite, right, jump over the Nine Mile base in 1941.
EARL COOLEY COLLECTION

Foreman Art Cochran, third from left, inspecting equipment and assisting the eight-man crew to ready for practice jumps. Camp Menard, June 1944.
PHIL STANLEY PHOTO

Checking chutes at Camp Menard, 1944.

at the start of the season. The others had gone to military service or were instructing parachute rigging with the CAA or the Army and the Navy. Only a few of the 33 recruits had any firefighting experience. Again all operations were concentrated at the Ninemile base. It was now more important than ever to protect the forests as part of the war effort.

Considering the reduced crew and lack of equipment, it was a lucky break the 1942 fire season was not critical.

In the Spring of 1943 the jump program looked like it was on hold for the duration of the war. Again, only five experienced jumpers returned and efforts at recruiting had produced only four candidates, none of whom could pass the military physical. The Civilian Public Service (CPS) Program came along and kept the smokejumping program going until the end of the war.

At the Nine Mile base, training facilities were greatly upgraded to take care of the new CPS volunteers. Equipment became plentiful although improvisations and substitution were still the rule of the day.

Again the summer produced few fires, but September and October were dry months and the jumpers were the only fire crews, air or ground, available for firefighting.

Expansion of the CPS Program continued in 1944. All training and refresher courses were in the Northern Region and were conducted at the new base camp, Camp Menard, below Stoney Creek. Then the crews were sent out to the Regional bases. Fire activity was below normal for the year but over 100 fire jumps were recorded. Large numbers of jumpers were dropped on larger fires. This proved that the program could handle this type of firefighting. For the first time, the program was integrated into the normal regional fire management plans and some forests even reduced their regular ground crews, depending on the smokejumpers to control the fires. This produced greater economy and use of regular crews.

The final year of the war, 1945, saw increased activity and development of the smokejumper program. A base was reactivated at Winthrop, Washington, and the total number of jumpers was increased to 220, most of them CPS men. This would be the worst fire season in the West since the inception of the jump

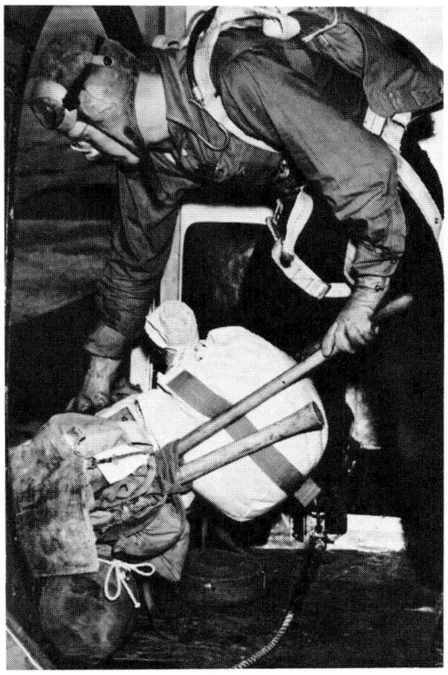

Bill Wood getting ready to push a fire pack out to a jumper, 1945.

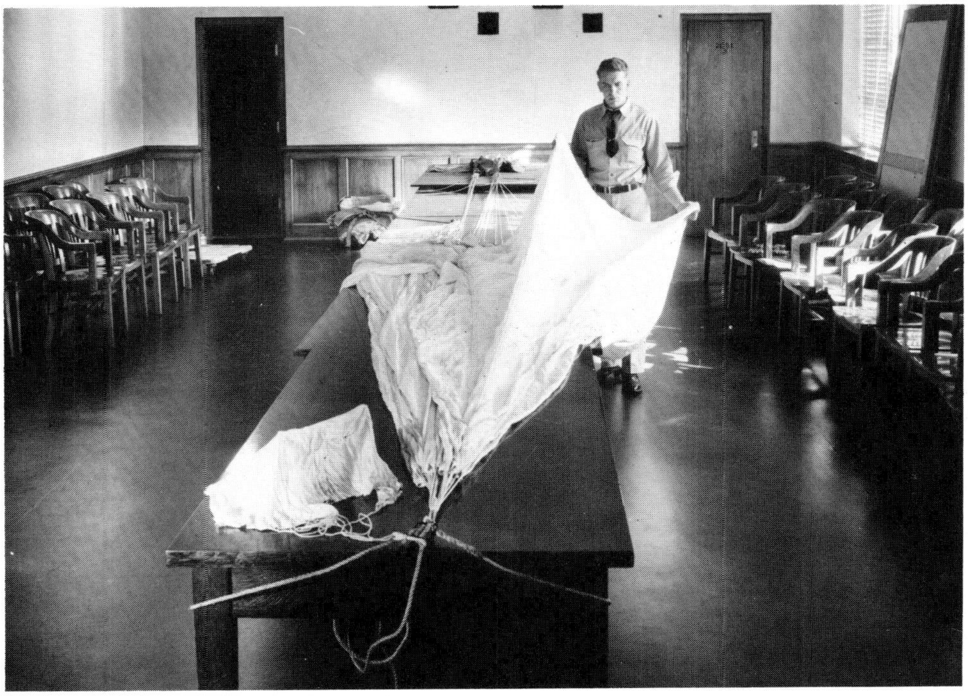

Chet Derry packs an Eagle parachute in 1940, in a conference room of the Federal Building in Missoula.
K.D. SWAN PHOTO

program in 1940. A total of 1,236 individual jumps were made on 269 fires. Jumps were also made into Glacier and Yellowstone National Parks, Indian reservations, private timber association lands and even one just over the border into Canada. This year truly demonstrated the effective use of smokejumpers to combat forest fires. The savings in dollars was considerable and the number of acres burned was reduced sharply considering the severity of the fire season.

The Northern Region set up an experimental "air control area" along the Continental Divide in Montana to be handled mainly by aircraft and jumpers to the exclusion of ground forces. The smokejumper program, started just before the war, had served the war effort well. Now it was time to return to more peaceful pursuits.

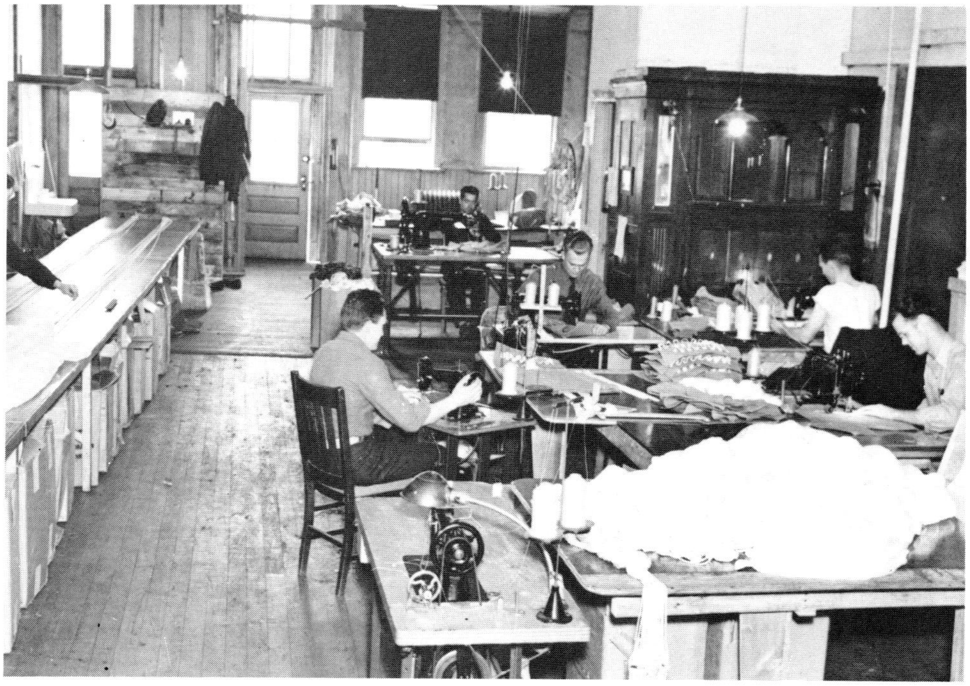

During the war a parachute loft was set up in the Park Hotel in downtown Missoula.
PHIL STANLEY PHOTO

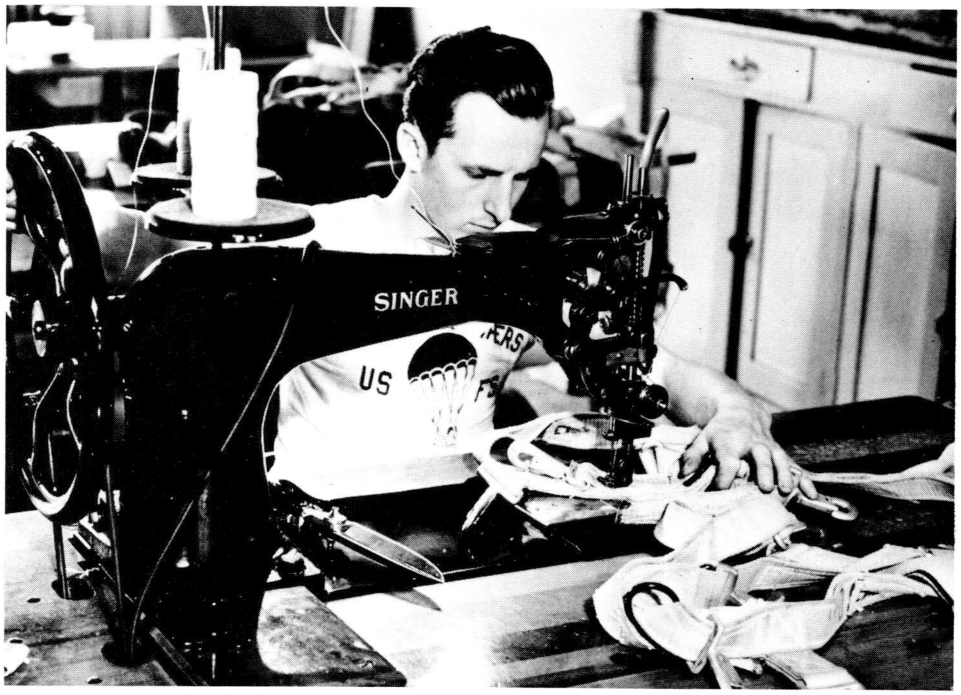

Camp Menard training camp. Parachute loft in background with highline poles and cables beyond jump training tower in right center. Large and small mock-up is in extreme right. High and low ramps are on the left side of picture, 1944.
PHIL STANLEY COLLECTION

Roy Mattson, left, and Earl Cooley, right, at the Moose Creek parachute loft, Nez-perce National Forest in Idaho, 1940. This was the first Forest Service loft established in the United States.
EARL COOLEY COLLECTION

Civilian Public Service (CPS Conscientious Objectors) smokejumpers at Camp Paxson on Seeley Lake, Lolo National Forest, Mont., 1943. They were housed at Camp Paxson while training at Seeley Lake. PHIL STANLEY COLLECTION

Camp sign at the Ninemile base, 1940s.

Jumper exits jump tower dummy door in excellent position. Rope from single-tree and risers is shown running through boom pulley, mast pulley and down mast to dead-en anchor. Safety man is located at this point to inspect rope and anchor after every jump and to pull single-tree back up within the grasp of the trainer.

Static line hookup, first used by the jumpers in 1941.

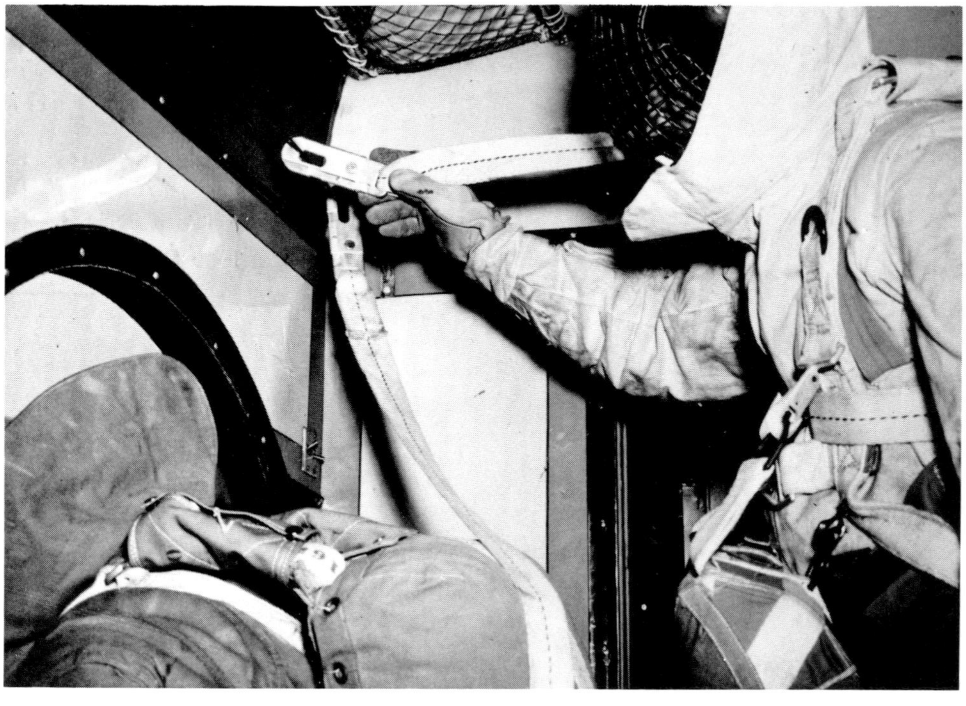

Foundation remains at Camp Menard.

Present-day picnic grounds. Site of Camp Menard, Nine Mile.

Site of loft building and jump tower training area, Camp Menard.

Present-day view of the Nine Mile airstrip.

Loading supplies at the Libby, Montana, airport to be dropped to a high, remote camp on the Bull Lake Fire, Kootenai National Forest, 1939.
K.D. SWAN PHOTO

Johnson Flying Service, based in Missoula, pioneered in mountain flying in Region One and in the development of the smokejumpers program in the late 1930s and 1940s. The service was virtually the region's air arm in this time period.

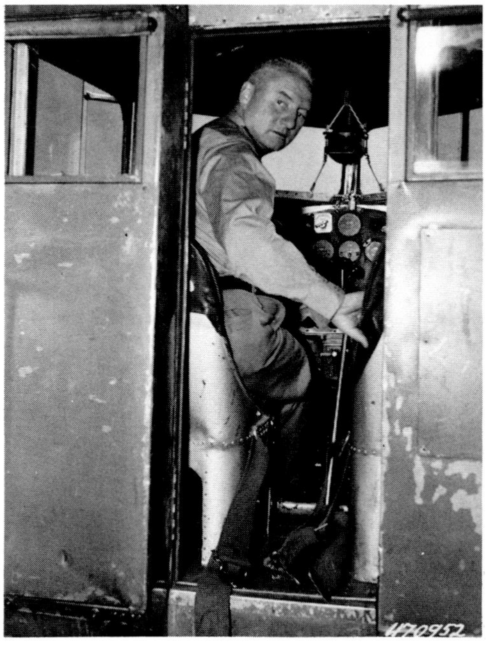

Bob Johnson, pioneer pilot and a founder of Johnson Flying Service, in the cockpit of a Ford Tri-Motor, September, 1945.
PHIL STANLEY COLLECTION

CHAPTER THREE:

Lookouts

In 1932 the Jackknife Lookout was in service in the Coeur d'Alene National Forest. It is another example of more advanced design features.

Firefinders were frequently located on high windswept peaks. Here, it's cold in the shade (if there is any), hot in the sun, and a tough climb to get there.

A wind-swept ridge, a rock slab, three posts, a map board and firefinder in 1921 made up the Roman Nose Lookout in the old Pend Oreille National Forest in Idaho.

Steve Bryant, putting the "Fire-Finder" on a smoke in 1927 from the Van Camp Lookout, Lochsa Ranger District, Selway National Forest.

The Crow's Nest Lookout Tower under construction in the Clearwater National Forest.

The reconstructed Crow's Nest ready for service as a lookout.

Firefinders were frequently located on high wind swept peaks. Here it's cold in the shade, (if there is any) hot in the sun, and a tough climb to get there.

Early Lookouts

In 1930 this "lookout tree" was in use on Hells Half Acre Mountain in the Nez Perce National Forest. Imagine spending your summer living in a tent and climbing a tree every day to look for fires.

In 1923, the Sliderock Mountain Lookout, administered by the Lewis and Clark National Forest, had the barest necessities for a lookout—a fire-finder and map board.

A wooden tower, that looks like it may have swayed in the wind, was soon added on the top of Sliderock Mountain.

"Primary lookouts" were permanent installations located at strategic points throughout the forests. They were manned from the time snow left in the Spring until it returned in the Fall. But when fire danger became extreme, "secondary lookouts" like this one near Whiskey Jack Ranger Station in the Lolo National Forest were also used until conditions improved.

This is a small shelter constructed near the Slide Rock Mountain Lookout. Note the lightning protectors over the roof.

If you can't see far enough, you can always build a very long ladder and lean it up against a very tall tree. This "ladder lookout" was erected in the Lolo National Forest.

The Gold Peak Lookout, originally constructed by one man working alone in the old Missoula National Forest, (now the Lolo) was condemned for use on the day this picture was taken—August 22, 1920.

In 1923 the "lookout tree" on Mount George in the Nez Perce had to be one of the most uncomfortable lookouts in the entire National Forest System.

Lookout Sorenson on duty at the Jim Creek Lookout Station in the Flathead National Forest, August, 1923.

In 1923, the Red Meadows Lookout on the old Blackfeet National Forest took advantage of a live tree for support and even some shade. It would be scary duty, however, to occupy this lookout on such a bare ridge during a lightning storm.

Cold Peak lookout tower and guard cabin in 1923 in the Missoula (now Lolo) National Forest.

Putting up block line. The ladder legs, cut from a nearby lodgepole pine thicket, were skidded to this location by saddle horses.

Paul Wright, Alternate Ranger, Lochsa Ranger District and Harry Cummings, Lookout Fireman, "sight in" the map board from the Crow's Nest Lookout in the Clearwater National Forest. (Harry Cummings, before his retirement in the mid-1980s, was one of the best Class 1 Fire Bosses in the entire U.S. Forest Service).

In 1932, Roy Perkins rests for a moment on a support of the East Dennis Lookout, Lochsa Ranger District, Clearwater National Forest.

Glenn McHone hanging telephone line to 62 Lookout in 1926. By 1932 there were few trees left standing in this burned area. Poles were erected to hold the telephone line in 1932 from the Moose Creek Ranger Station to the 62 Lookout.

Setting a phone line in the 1920s.

The old Iris Peak lookout in 1930. The replacement lookout is under construction on the left.

The Sunset Peak lookout on duty sighting a smoke through the firefinder in the Panhandle National Forests.

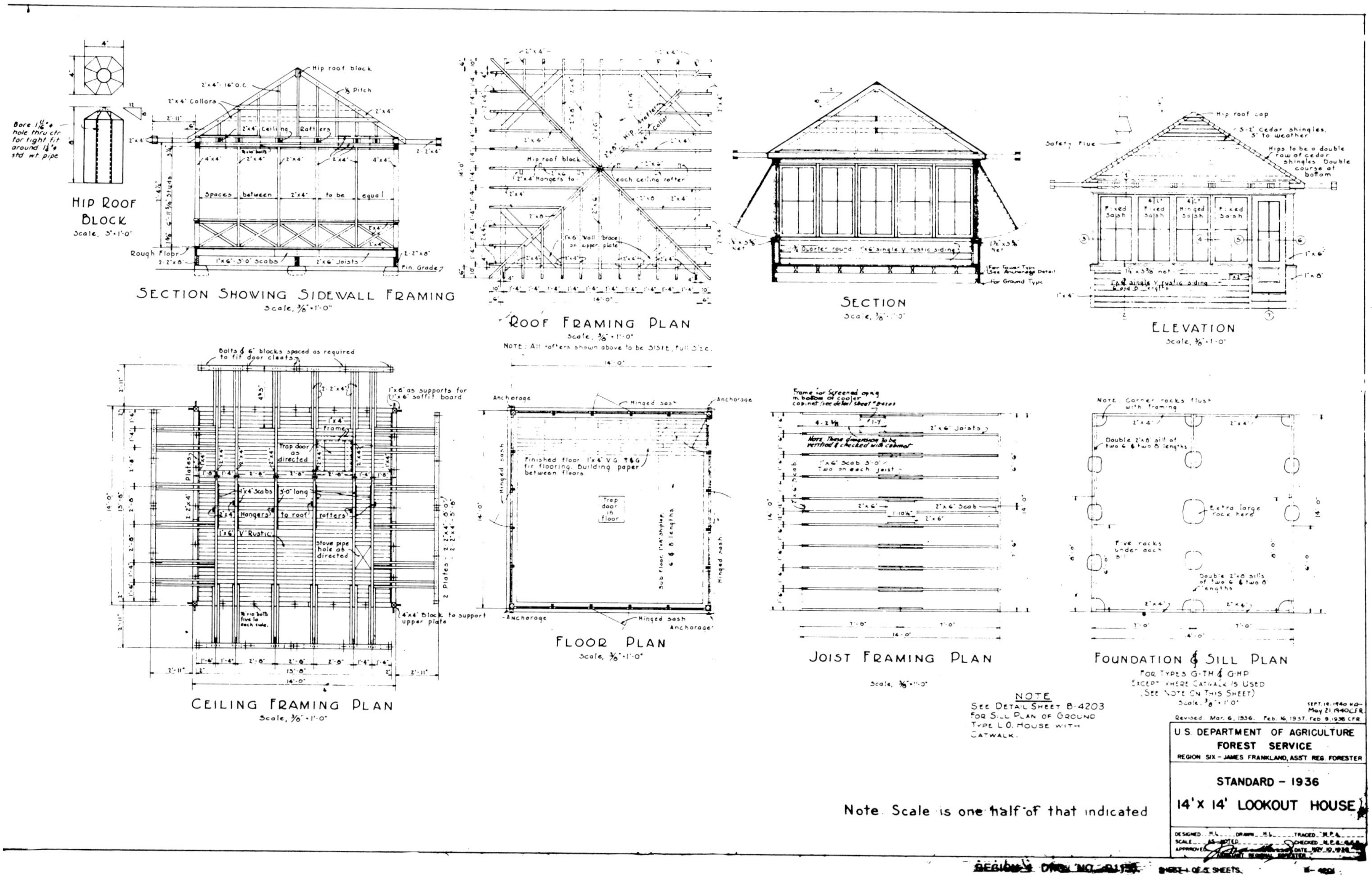
HIP ROOF BLOCK
Scale, 3"=1'-0"
Bore 1 1/16" ⌀ hole thru ctr for tight fit around 1/8" ⌀ std wt pipe
SECTION SHOWING SIDEWALL FRAMING
Scale, 3/8"=1'-0"
Hip roof block
1/3 Pitch
2"x4" Collars
Ceiling
Rafters
Spaces between 2"x4" to be equal
Rough floor
1"x6" 3'-0" Scabs
2"x6" Joists
Pin Grade
ROOF FRAMING PLAN
Scale, 3/8"=1'-0"
NOTE: All rafters shown above to be S1S1E, full size.
Hip roof block
2"x4" Hangers to each ceiling rafter
1"x6" Wall brace on upper plate
SECTION
Scale, 3/8"=1'-0"
3/4 Quarter round
1"x6" single V rustic siding
ELEVATION
Scale, 3/8"=1'-0"
Hip roof cap
5-2" Cedar shingles, 5" to weather
Hips to be a double row of cedar shingles Double course at bottom
Safety flue
Fixed Sash
Hinged Sash
CEILING FRAMING PLAN
Scale, 3/8"=1'-0"
Bolts & 6" blocks spaced as required to fit door cleats
1"x6" as supports for 1"x6" soffit board
1"x4" Frame
Trap door as directed
1"x4" Scabs 3'-0" long
2"x4" Hangers to roof rafters
1"x6" V Rustic
Stove pipe hole as directed
4"x4" Block to support upper plate
2 Plates
FLOOR PLAN
Scale, 3/8"=1'-0"
Anchorage
Hinged sash
Finished floor 1"x4" V.G. T&G fir flooring, building paper between floors
Trap door in floor
Sub floor 1"x6" shiplap, 6 & 8 lengths
JOIST FRAMING PLAN
Scale, 3/8"=1'-0"
Frame for Screened opng in bottom of cooler cabinet (see detail sheet B-4202)
2"x6" Joists
Note: These dimensions to be verified & checked with cabinet
1"x6" Scab 3'-0" Two on each joist
2"x6" Scab
FOUNDATION & SILL PLAN
For Types G-TH & G-HP Except where Catwalk is Used (See Note on this Sheet)
Scale, 3/8"=1'-0"
Note: Corner rocks flush with framing
Double 2"x8" sill of two 6 & two 8 lengths
Extra large rock here
Five rocks under each sill
Double 2"x8" sills of two 6 & two 8 lengths
NOTE
See Detail Sheet B-4203 for Sill Plan of Ground Type L.O. House with Catwalk.
Revised Mar. 6, 1936. Feb. 16, 1937. Feb. 9, 1938 CFR
Sept. 14, 1940 HG
May 21, 1940 CFR
U.S. DEPARTMENT OF AGRICULTURE
FOREST SERVICE
REGION SIX - JAMES FRANKLAND, ASST. REG. FORESTER
STANDARD - 1936
14'x 14' LOOKOUT HOUSE
Note Scale is one half of that indicated
DESIGNED H.L.
DRAWN H.L.
TRACED
SCALE AS NOTED
CHECKED
APPROVED
SHEET 1 OF 5 SHEETS
B-4201

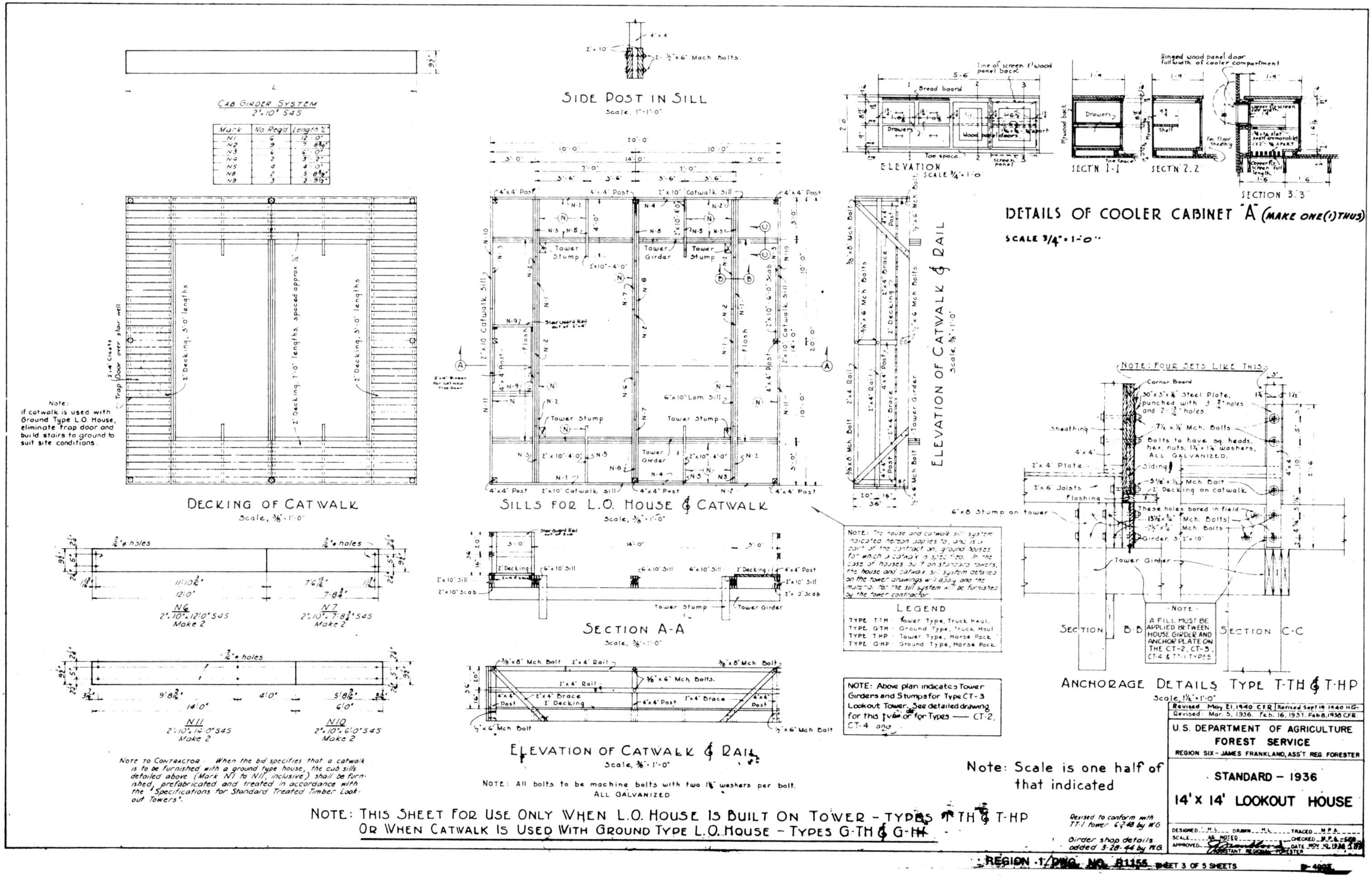
CAB GIRDER SYSTEM
2"x10" S4S
SIDE POST IN SILL
Scale, 1"=1'-0"
DECKING OF CATWALK
Scale, 3/8"=1'-0"
SILLS FOR L.O. HOUSE & CATWALK
Scale, 3/8"=1'-0"
SECTION A-A
Scale, 3/8"=1'-0"
ELEVATION OF CATWALK & RAIL
Scale, 3/8"=1'-0"
DETAILS OF COOLER CABINET "A" (MAKE ONE (1) THUS)
SCALE 3/4"=1'-0"
ANCHORAGE DETAILS TYPE T-TH & T-HP
Scale 1½"=1'-0"
Note: If catwalk is used with Ground Type L.O. House, eliminate trap door and build stairs to ground to suit site conditions.
Note to Contractor: When the bid specifies that a catwalk is to be furnished with a ground type house, the cab sills detailed above (Mark N1 to N11, inclusive) shall be furnished, prefabricated and treated in accordance with the "Specifications for Standard Treated Timber Lookout Towers".
NOTE: All bolts to be machine bolts with two 1⅛" washers per bolt. ALL GALVANIZED
LEGEND
TYPE T-TH Tower Type, Truck Haul.
TYPE G-TH Ground Type, Truck Haul
TYPE T-HP Tower Type, Horse Pack
TYPE G-HP Ground Type, Horse Pack
NOTE: Above plan indicates Tower Girders and Stumps for Type CT-3 Lookout Tower. See detailed drawing for this type or for Types — CT-2, CT-4 and
NOTE: A FILL MUST BE APPLIED BETWEEN HOUSE GIRDER AND ANCHOR PLATE ON THE CT-2, CT-3, CT-4 & T-1 TYPES
Note: Scale is one half of that indicated
NOTE: THIS SHEET FOR USE ONLY WHEN L.O. HOUSE IS BUILT ON TOWER - TYPES T-TH & T-HP
OR WHEN CATWALK IS USED WITH GROUND TYPE L.O. HOUSE - TYPES G-TH & G-HP
Revised May 21, 1940 CFR | Revised Sept 19, 1940 HG
Revised: Mar. 5, 1936. Feb. 16, 1937. Feb. 8, 1938 CFR
U.S. DEPARTMENT OF AGRICULTURE
FOREST SERVICE
REGION SIX - JAMES FRANKLAND, ASS'T REG. FORESTER
STANDARD - 1936
14' x 14' LOOKOUT HOUSE
Revised to conform with TT-1 tower 6-2-40 by HG
Girder shop details added 3-28-44 by HG
REGION 1 DWG. NO. B1155 SHEET 3 OF 5 SHEETS

Clyde Fickes, an early day Ranger, (who was recognized in later years as a designer of lookouts), with his horse ready for the trail.

The Striped Peak Lookout Coeur d'Alene National Forest in 1924 was the first lookout designed by Clyde Fickes. After this, attached cupola's were removed from the design.

A lookout plotting the location of a smoke he has spotted from the Colt Mountain Lookout, Couer d'Alene National Forest.

The Black Pine Lookout Tower was constructed in 1939 in the Deerlodge National Forest and this picture was taken in 1940. This lookout is now on display in the Visitor Center of the Forest Service Aerial Fire Depot near Missoula, Montana.

Mrs. Hollis Stritch looks out the window of the West Fork Butte Lookout in the Lolo National Forest. This lookout has "all the comforts of home"– especially when compared to the ladders or tree lookouts that served other locations.

In areas that do not have mountains, tall, steel lookout towers were needed to adequately see the surrounding area. In 1928, this tower looked over the Long Pine Hills, Custer National Forest.

A very comfortable structure, the Lindrosa Hill Lookout served the area near Mullan, Idaho, in 1938. A picnic area was also maintained by the Forest Service on this site.

What can a lookout look like in winter? The above snow plastered structure is Vista House on the top of Mt. Spokane. It was built as a fire lookout in 1913.

The Blackfoot Forest Protective Association maintained this lookout in 1944 on Sunset Point near the Lolo National Forest.

The most important job of the Forest Service "look-out" was to spot fires while they produced only a faint wisp of smoke. Temperatures on most nights in the High Country drop below freezing—the stove is needed for warmth as well as cooking. Canned and dry food provided sustenance-fresh vegetables and meat was a rare treat. Clean windows were very important so smokes were not missed. This lookout, from Sliderock Mountain, Deerlodge National Forest, is currently on display at the Historic Museum At Fort Missoula, Missoula, Montana.

There was a stove with its all important coffee pot; shelves for storing condiments, canned and dry food, and a bunk. This lookout, Black Pine, from the Philipsburg Ranger District, Deerlodge National Forest, is currently on display in the Forest Service Visitor Center at the Aerial Fire Depot, U.S. Highway 10 West, Missoula, Montana. No longer needed, it was dissembled and moved by Smokejumpers to the Aerial Fire Depot in 1978. It was reassembled and put on display in 1979. Had it not been moved, it would have burned to the ground in 1988 when the Combination Fire swept through this area.

There was an acetylene light on the ceiling; a fire finder, the most important tool in the lookout, in the center of the floor; stool insulated against lightning, a bed with insulated legs; and the all important telephone for reporting smokes.

CHAPTER FOUR:

Ninemile Ranger Station

Ninemile Remount and Winter Range

by Lee Hames, Missoula Montana.

The author of this article, Lee Hames, is a descendent of early day settlers in the Bitterroot Valley of Western Montana and spent part of his youth on the Fort Owen ranch near Stevensville, Montana. For several years after high school, Hames' spent his summers as a cowboy and winters in area logging camps. He also participated in rodeos and spent short periods working for the U.S. Forest Service breaking and training horses and mules or working as a packer on fires. During World War II he served as a pack artillery officer in the China-Burma-India Theater and served a short time after the war at the American embassy in Chunking, China. He spent many years ranching, logging and engaged in the livestock hauling business in western Montana in addition to being active in the rodeo business around the west. He is now retired and lives in Lolo, Montana.

From the time that the Forest Service was established in 1905, pack animals were used to transport all types of equipment. The forests were divided into districts that operated as sort of separate little islands. Trails and telephone lines were built to high peaks where observers spent their summers watching for forest fires. Those trails and telephone lines had to be maintained by crews of men who needed tents, beds and food, plus shovels, axes, picks, and sometimes, even dynamite.

Supplies to the men in the lookouts went in on a weekly basis on pack animals. Even the lookout itself, a 12- to 16-foot square building with glass windows on four sides, was packed in. Observers stayed in these lookouts from June to September each year, locating fires and giving directions to get to them. During the actual fighting of fires, animals, supplies and equipment were trucked as near to the fire as possible and then the supplies were packed by horse or mule to wherever they were needed. At first, mostly horses were used, but as years went by the mules began to replace horses. The mules walked straighter, carried heavier loads, did not bump loads against trees and in most ways were preferred by the packers. Most districts kept a few head of stock but depended upon renting local stock during fire season.

The year of 1929 was the start of the drought years of the 1930s. Spring came early without any spring rains, followed by a very long, hot, dry summer. Our family was ranching 10 miles east of Stevensville on the old Grist Mill Ranch. Burnt Fork Creek barely flowed from the canyon.

About the middle of haying season that summer, I had my first contact with the Forest Service. A forest fire had erupted on a tributary of Burnt Fork Creek known as Gold Creek. The fire was about eight miles southeast of our buildings near Burnt Fork Lake. It was in steep, rocky and heavily timbered country, with no roads and very few trails. A ranger called on my dad, inquiring about the possibility of using our corrals and sheds for a fire base camp. They also needed to rent pack and saddle horses plus riding and pack equipment. The Forest Service did not have any stock of their own to place on the fire. Dad rented them a good set of corrals and a large sheep shed. The Forest Service more or less commandeered all of dad's horses except the teams dad had in the hay fields. Some additional stock was rented from neighboring ranchers. Two local cowboys, Horace Pope and Felix Phelps, were hired to do the packing. Altogether the Forest Service must have rented about 30 head of stock. When the trucks arrived with the fire fighting equipment, dad and his haying crew helped the packers saddle and load the supplies. They put cowbells on two mares and turned the entire herd loose. They had to go several miles up the country road before coming to a trail that would lead them in the direction of the fire. I had a small sorrel saddle mare that was apt to buck if anything scared her. Because she was small, the packers put a light load on her. The lightest load they had were boxes of tin dishes. When the mare heard the dishes rattle, she exploded and bucked her packs off, scattering dishes. We helped gather the dishes, caught the mare, and the packer tied the load on more securely. Conditions like this were typical of all forest districts throughout the west. With the cooperation of the public, the Forest Service did whatever needed to be done during the fire season.

To use the words of Clyde Fickes, from the Northern Region, the year of 1929 turned out to be a total mess and disaster. Enough good men, stock

An early picture of the buildings at the Nine-mile Remount Depot after their construction by the CCC.

A pack string "lining out" from the Ninemile Remount Depot. This string of nine mules can pack over a ton of supplies on their backs. The term "Remount" was borrowed from the U.S. Cavalry. This picture was taken during its first years when 30 to 35 employees worked hard to turn a run-down leased ranch into a good working unit.

and equipment just could not be found in a bad season. Something had to be done to provide a better coordinated effort in fighting forest fires.

During the winter of 1929-30 Maj. Evan Kelley became regional forester for the Northern Region. His title of major was earned as a cavalry officer during World War I. Major Kelley's proposal was to establish a remount service similar to that of the U.S. Army. A centrally located remount would provide the Forest Service with the following:

1. A reservoir or pool of experienced packers and pack animals that the districts could draw from when needed.

2. Development of the best type of riding and packing animals for use in the forests.

3. Adequate means of transportation for men, stock and equipment on highways.

4. A Remount Center to be used as a training place for both men and stock. Testing would be done to develop the best kind of fire fighting equipment.

5. Standardization and improvement of methods of packing and care of stock and equipment.

The Ninemile Remount Depot buildings were designed by Missoula, Mont., architect Bill Fox in a distinctive Cape Cod style.

Clyde Fickes, left, who originally proposed establishing a remount depot and administered the facility as it started, looks over facilities under construction.

The vision of the Remount started with the belief that the ranger districts would continue to maintain some pack stock within their district and would only use the Remount stock as a pool to draw upon in emergencies. For this reason the Forest Service did not look for a very large ranch upon which to establish the Remount. They located a place, the Allen Ranch in the Ninemile Valley, 30 miles west of Missoula, which they thought would serve their purpose. The ranch had good irrigated hay meadows, some pasture, was not too far from Highway 10, and there were sidings on two railroads within a few miles. They leased the Allen Ranch from 1931 to 1934 before purchasing it. Little did they realize how quickly they would outgrow the one-square-mile that was the Allen Ranch. The ranch was very short of pasture during the summers when lots of stock had to be held there. The ranch provided no winter range at all because of the deep snows in the valley.

The Forest Service built some corrals and fenced an area for additional pasture not too many miles from the Remount, south of the Clark Fork River, between Albert and Tank creeks. They also used the Bald Mountain at Keystone, west of Superior, for some winter pasture.

At this point in the operation, a decision was made to no longer leave any stock in the ranger districts during the winter months, so a search for a winter range began. Two ranches, the Bud Bruns and Paul Foote places, down in what is known as the Big Draw, between Nirada and Elmo, were leased. This is an area of mild winters with very little snow. It was an ideal place for stock to winter without being fed any hay. The drawback to the Big Draw location was that it was about 100 miles from Ninemile.

In 1933 the U.S. Government established a large CCC Camp at the Remount. The CCC's were made up of eastern city boys who had hardly ever been off a sidewalk let alone touched a mule. Bossed by Forest Service employees, the CCC's cut poles, built miles of jack-leg fences, cleared land, built corrals and did it all with mule teams. A day never went by without some runaways. One old timer said that if you wanted to take a picture of a mule team running away, all you had to do was point your camera in any direction during daylight hours.

New white buildings, surrounded by shrubs and flowers replaced the old farm ones. When the Remount was completely rebuilt, it looked like a cross between a Kentucky breeding farm and an Army remount post.

In 1935 an opportunity arose to lease several blocks of land on the north side of the Flathead River between Perma and Dixon. This area was only about

A pack string crossing the Lochsa River near Jerry Johnson Hot Springs.

20 miles distant from the Remount. An easy day's horseback ride up the Ninemile Valley over a small divide called Sego Pass would put you on the Flathead River. The big drawback was that there was not a bridge across the river at this point. A few miles upriver from these ranches, there had been an old ferry known as Sloans Ferry. The old ferry was abandoned when replaced by a bridge. The Forest Service moved the old ferry downstream and used it to cross the river to some corrals and a few buildings that became known as the Ferry Basin Ranch. East of Ferry Basin was a place called the Pete McDonald Ranch. To the north of these two ranches was a very large acreage referred to as the Mahoney Estate lands. In early days the Mahoney land was the wintering and lambing area for a very large sheep operation. North of the Mahoney land was a smaller ranch known as the Gustine Place. If you crossed the Flathead River at the Perma Bridge, you could come upriver to get to the Mahoney land. It was at the end of this road that the Forest Service built the winter range headquarters, with homes, a blacksmith shop, hay sheds and acres of corrals. Corrals and horseshoeing shops were also built at Ferry Basin. The Gustine

A view of winter range administrative structures in the Big Draw of the Perma Area about 90 miles north of Missoula, Mont.

Ranch was always used as a place for brood mares. Some of the stallions were kept there at different times, but during the last few years before the brood mare dispersal sale, it was used mostly to raise mules. They had a long row of covered box stalls which would house seven jacks.

In 1948, I went to the Mule Mare Dispersion Sale at the Gustine Ranch, where they sold about 90 brood mares that were due to have mule colts that spring. Howard Raser and Tad Brown bought most of the 1,400 pound mares for about three cents per pound. The days of the Forest Service mule raising came to an end. The jacks had already been disposed of, and there was really no market for the mares. Everyone was mechanizing and no one wanted any mules. During the 1930s, at about the same time that the Forest Service was establishing the Remount, the ranchers and farmers of America were beginning to replace horses and mules with tractors and crawlers.

In the large grain-producing areas of Montana, Alberta and the Palouse country of Idaho and Washington, however, large horse-drawn grain combines were used. In these areas, a certain type of work animal had been developed. Since the big combines worked as many as 32 head of horses in one hitch, and had all of the gears ground-driven by a large bullwheel that turned as the horses moved forward, horses with considerable weight and a very fast walk were required. To get this type of animal, large draft mares were bred to thoroughbred or standardbred stallions. Mature horses of this cross produced compact, blocky animals that weighed 1,400 to 1,500 pounds and were about 15 hands high. Mares of this size were ideal mothers to produce the 1,100 to 1,200 pound mules that the Forest Service wanted to raise. Cap Evans and Walter Peery went to the Palouse country in 1935 to buy 100 head of these crossbred mares. The mares were shipped by rail to Perma and driven to the Gustine place. Two mammoth jacks were moved from Ninemile to the Gustine place. Breeding jacks to mares is a rather difficult job and the first year's mule colt crop was disappointingly small. Additional jacks were purchased, breeding methods improved, and eventually mule production got underway. Walter Peery told me that in the years that he worked for the Remount, the largest number of mule colts from these mares that he could remember was about 90.

When Walter Peery went to work at the Remount, he was considered to be one of the better horsemen in this area. He took a black stallion called Laddie with him. Laddie was descended from Daly Ranch stock but was not a registered horse. My dad and Walter bought some horses from the Daly Ranch that did a lot of pacing so I believe the horse was probably part Standardbred. Walter thought the horse may have been part American Saddler. The Forest Service raised some colts from Laddie which were very good, useful horses. Jake Williams rode Laddie a lot and really liked him. Partly because of Laddie, the Forest Service decided to use American Saddler stallions. The first Saddler stallion they purchased was Grand Menard. He was bought at an auction sale in Nebraska for $2,500, but he was raised in Missouri. He was a big, beautiful sorrel but did not seem to sire very good colts. Most of his colts seemed to be stubborn, sullen and slow learners. I broke a couple of colts from the horse

Stock in corrals at the Perma Winter Range. In the 1940s the Winter Range held over 250 mares and four to six jacks. This herd produced over 200 mule colts each year.

Shoeing shed at the Ferry Basin winter Range.

Photos from Lee Hames, 1940

The Ninemile Remount Depot in 1940.

Corrals at the Ninemile Remount Depot in 1940.

Forest Service mule string in 1940 starting on the trail.

Forest Service mules in 1940 packed and ready to go.

Loading out for a fire in 1940 at the Ninemile Remount Depot.

Forest Service mules in 1940.

and was disappointed with them. They were slow, sullen and bad to strike. They also used a Saddler named Cascade Highland Fox that produced some fair colts. Two other American Saddler stallions, Hermes and Kiniksu Chief, were also used. Actually the Saddlebreds did not work out too well. They were too flighty, walked too fast for pulling a pack string, and were taller than the packers liked.

The Morgan horses were also used at different times. A sorrel Morgan horse, raised by Tom Adams, called Teddy Roosevelt, was at the Remount at one time. They acquired a Morgan from a Washington state man named Freddie Brown and that was what the stallion was always called. This horse was bad to buck and so were most of his colts. A horse sired by Freddie Brown was the sire of Trails End, the world's top saddle-bronc in 1959.

When Henry Wallace, (secretary of agriculture at the time), visited the Remount in 1936, he brought with him some Hungarian Nonius horses. There were two stallions, Shorty and Kipling, and a half a dozen mares. They were mostly brown or light black, stocky, with good legs and feet and weighed around 1,100 pounds each. They seemed to be very tractable, mild-mannered and looked like an ideal dual-purpose (ride or pack) type of horse. They had a fair walk and looked as if they should be very useful mountain trail horses. The Hungarians used them as a dual-purpose military horse for both riding and driving. For some reason the Forest Service did not seem to think they were very good so did very little with them. When the Forest Service sold their breeding stock, Bud Riley, from Drummond, bought most of the Nonius horses. Later, Riley had a sale at which Charles Smith of Missoula bought the Nonius horses. Smith tells me that most of them are now gone. No one seemed to have been interested in perpetuating the breed.

In the fall, all of the stock would be moved to the winter range where they would be turned out to graze and rest for the winter. When they were rounded up in the spring, they would be corralled at either Ferry Basin or winter range headquarters. Both places had blacksmith shops where they put shoes on the stock before distributing them to the districts for the summer. The shoers worked in three man crews. One man did nothing but trim feet and nail shoes. One man worked on the forge and shaped all of the shoes. The third man had a long soft rope that he used for tying up the feet on the stock that objected to the shoeing. With several crews working they went through a lot of stock in a short time. The shoers did not waste any time. A good crew could shoe 20 to 30 head per day.

Several hundred head of horses and mules had to be shod and distributed to the districts in a few days time. If a horse or mule acted up, they put trip ropes on him and laid him on the ground. Shoes were nailed on while he was on his side and then the animal was allowed to regain his feet. Anything that went into the shed barefooted, came out walking on iron. During the summer months horse shoers traveled district to district, replacing shoes about every six weeks. Dave Pronovost was supposed to have shod over 30 head in one day. He was probably their top shoer.

The winter range was heavily infested with rattlesnakes. You could not walk very far in any direction without encountering one. The men who had families living there, kept their yards fenced with mesh wire to keep the snakes out. Small children were not allowed out of their yards unless they were with an adult. George Tyler, a buckaroo from the southern Idaho-northern Nevada country, came to work at the winter range. He was raised in rattlesnake country and didn't like them one bit. He said where he came from, they got rid of the snakes by turning hogs loose near their dens. George claimed the hogs

A mule loaded with lumber.

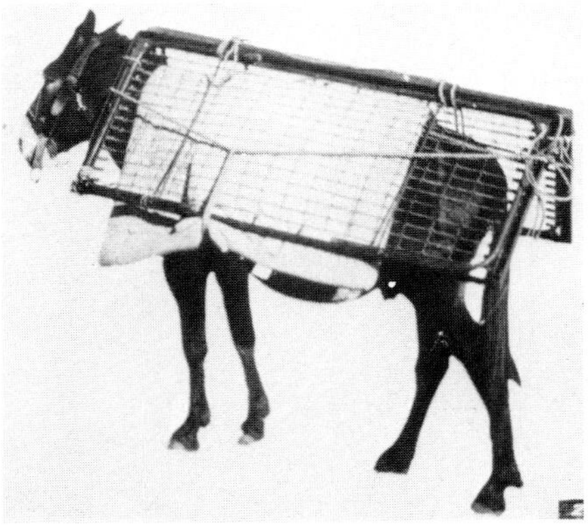

Packing cots.

When feed is scarce, you can always pack it in.

A convoy hauling mules to a trail head.

would catch, kill and eat all of the snakes. The winter range boss bought half a dozen 100 pound porkers and turned them loose around the buildings. Every so often you would see a hog dragging a snake across the corral or barnyard. It was not very long until you seldom saw any snakes around the buildings.

If snakes could laugh, they probably had the last laugh on George. He was breaking one of Freddie Brown's colts to ride (a big brown horse that was really bronky). On this day, George had spent several hours out on the range trying to take the snort out of the colt. Coming back to the buildings the colt was tired and had settled down, behaving himself. George was relaxed and was drowsing along in the saddle. As he rode by the hay shed, the hogs came squealing out and ran underneath the colt. Caught unawares, the colt bucked George off, breaking his wrist. George had to ride for several weeks with a cast on. While he was laid up, several of the other cowboys tried to ride the brown colt. The more times they tried to ride the horse, the harder he would buck. Roy King, a local rodeo producer, heard about the horse. Roy traded the Forest Service several well-broken saddle horses for the horse and put him in his bucking horse string. Roy named the horse "The Brown Bomber." A few years later Roy went out of the rodeo business and the Bomber was sold to Leo Cramer, one of the world's largest rodeo producers. The Bomber came to be one of the best known broncs in the Cramer string.

A pack saddle loaded with stakes and cable ready to go.

After the Forest Service disposed of their trucks, Sonny Lundwall and I were hired to truck the stock from the winter range to the various districts each spring. In the fall we trucked them back. Sonny had never been to the winter range before so I warned him about the rattlesnakes. When Sonny backed into the loading chute for the first time, he started to get out of the truck and heard the buzz of a snake. He looked down and there was a rattler lying in the cool dust of the road. He reached for a tire iron, stepped out and disposed of the snake. That was the only snake we saw during the years we did the hauling.

When the Forest Service first went into the mule business, herding the packed animals was the custom. An old mare, usually white or light-colored, would have a large cowbell strapped around her neck and would be led down the trail followed by the pack mules. The mules just seemed to love an old mare and would follow her almost anywhere. This was fine, as long as everything

Loading mules into trucks at the Ninemile Remount Depot in 1931.

A truck loaded with mules on its way to a fire in 1937.

Trucks at the Ninemile Remount Depot in 1940 ready to load.

Unloading mules.

went smoothly but if anything went wrong, look out, it could turn into bedlam. Some times a pack would turn under a mule's belly. The mule could stampede down through the herd, knocking mules right and left, dislocating packs, and creating all kinds of disorder. If a mule happened to step on one of those ground hornets nests that lay along the trail and get stung, away he would go. These little black hornets could cause more trouble to packers than anything else. A packer would rather meet a bear or a moose on the trail.

Controlling the animals around the fire area was also a problem, and some method of control had to be developed. From the first time that pack animals had ever been used in concentrated numbers, they had always been driven in a loose herd. The Army referred to this loosely driven pack method as a "pack train." The Forest Service tried leading three or four mules tied or pigtailed one behind the other in a string. This worked well but it took too many packers. They tried different lengths of strings up to and including as many as 16 head. The long strings worked all right on flat ground and straight trails but when

Secretary of Agriculture, Henry Wallace, planting a tree at the Ninemile Remount Station in 1936.

it came to steep, rocky trails with lots of switchbacks and turns, they were just too unwieldy to handle. They finally decided a string of nine pack mules and the packer's saddle horse worked the best. One man could easily saddle, load and handle this size of string on any type of terrain. A nine-mule string could deliver food, bedrolls, tens and tools to a 25-man crew fire-fighting outfit.

The next stage in their development was to find the best way to deliver the mules and equipment to the fires. In 1937, the Kenworth Truck Company, using a Greyhound bus chassis, built and delivered a mule-hauling truck to the Forest Service. It was one of a kind and had been built to the exact specifications of the Forest Service engineers. The truck was a cab-over-engine model, with a sleeper cab, dual axles and 10 wheels. The truck was designed to haul nine mules and the packer's riding horses, all of the packsaddles and riding saddles, and a complete 25-man crew pack outfit. When the truck went as close to the fire as it could, the string could be unloaded from the truck on a ramp, packed out and moved to the fire. The Forest Service bought additional Kenworth bus chassis and built several more units in their own shops. Large trailers with the same carrying capacity were designed to be pulled behind these trucks. This made it possible, roads permitting, to deliver two strings at a time to a fire. The trucks had six-cylinder, Hall-Scott gas engines that burned a gallon of fuel every two or three miles, depending upon road conditions. Buster Dickerman drove one of these trucks to a fire in Yellowstone Park one time. He said he had lots of power. He could pass everything on the road but a gas station. He was hauling two strings and their packs, so was heavily loaded. Buster claims that he spent about as much time looking for gas stations as he did driving.

The Forest Service goal of standardizing their equipment meant that they must first decide which type of packsaddle to adopt. The oldest and most widely used packsaddle in the world is the old wooden sawbuck. When the Forest Service was renting both stock and saddles that was usually the only saddle that they could get from the ranchers. It was a good saddle but required a lot of rope plus a lot of experience and skill in using the various hitches for different kinds of loads. Another saddle they experimented with was the Spanish aparejo. The aparejo looks like a leather mattress in a rope frame. Around the edges are holes through which ropes can be threaded to lace on loads. It was all right for some loads but not very satisfactory as an all-purpose packsaddle.

From 1930 until 1953, the Ninemile Remount Depot provided experienced packers and pack animals for fighting fires and for backcountry work projects throughout the vast roadless areas of the Northern Rockies. Modeled after U.S. Army Cavalry Remount Depots which supplied fresh horses to troops, the Ninemile Remount Depot is listed on the National Register of Historic Places for its unique role as a Forest Service Remount Depot and for its distinct Cape Cod style architecture.

The packsaddle finally adopted by the Forest Service was the Decker. A saddle similar to the Decker was originally patented by a man named Moore and was used by the U.S. Army in the 1860s. This saddle was discarded by the Army in favor of the McClellan which could be used for both riding and packing. A crude packsaddle similar to the Moore was made by two Idaho packers named Decker and McDaniels. A Selway packer, Al Robinette, using almost the same pattern, designed the present saddle, the main difference being that Robinette made each saddle for the individual back of each mule. The same saddle was always used on the same mule. The Remount had their own leather shop and every winter some of the packers spent their hours in the shop repairing saddles and equipment.

The Decker was a simple saddle to make, requiring just hand labor and time. Two pieces of cottonwood planks were shaped by hand to conform with the shape of the mule's back. These two wooden bars were fastened together with two iron arches that held the bars in the proper shape. Leather riggings and cinches were attached to the bars to keep the saddle on the mule. Over these arches, bars and rigging a stuffed canvas, called a half-breed, was placed to support the loads. A simple sling hitch was used to pack one load on each side. If a mule bumped his load into a tree, the load would swing and return to its original position. The saddle was light and required very little maintenance. All types of cargo, barrels, lumber, doors, windows, camp gear, shovels, axes, picks, hay and grain have been packed on Deckers. Every saddle was marked and numbered with the name and number of the mule that it fit, and as long as a mule or saddle remained in service, they always stayed together.

During World War II, the Forest Service sent Bill Bell to the Mountain Warfare Training Center at Camp Carson, Colo., to demonstrate the use of the Decker. At that time the Army was forming the Mountain Infantry and Pack Artillery for use in Italy and Burma and was trying out every type of saddle that they could find. They wanted a saddle that could be used to transport both cargo and military weapons, but the Decker could not be modified to support heavy artillery gun loads. The Army settled on a heavy packsaddle called a Phillips that was awkward but could be used for both guns and cargo.

The development of the Smokejump Center in Missoula sounded the death knell of the Remount. Logging and the use of large crawler tractors had ex-

Putting up hay at the Ninemile Remount Depot.

A packers contest at the Ninemile Remount Depot in 1931.

tended roads far back into the mountains. There would not be very much use for the mule any longer; men and equipment could be air-dropped into fires in far less time. The use of the parachute and supply plane had been developed to a fine art during World War II. Supplies were air-dropped to men and mules in the jungles of Burma, miles and miles from the nearest roads. They even tried to air-drop some replacement mules, without success. The Remount was considered to have served its purpose, and to was determined that it should be dismantled and replaced with the Smokejump Center.

The first part of the Remount service to fall under the axe was the breeding program. In 1948, they held two sales, one at Ninemile for the mares and stallions in the saddle horse program and one at the Gustine Ranch where the mares that produced the mules were sold. Some of the horses sold fairly well but some barely brought canner prices.

At about this same time the U.S Army also decided to do away with its Remount horse program. At Fort Robinson, Neb., the Army for many years had maintained a thoroughbred breeding farm to produce its own stallions. Stallions from this program were placed with ranchers throughout the country for the purpose of encouraging the raising of horses suitable for cavalry use. In 1950, off chance, a thoroughbred stallion and about a dozen mares were sent to the Ninemile Remount from Fort Robinson. They remained at the Remount until 1956 when they were sold at auction. This was the last of the horse breeding.

The Remount at Ninemile ceased to exist in 1953. One or two pack strings were to be kept there during the summer for use in the Ninemile District or they could be transported to another district when needed. John Christianson was the last packer that I knew who was stationed at Ninemile. The other ranger districts continued to keep a string in their district. During the winter months, all of the strings were taken to the winter range at Perma. The following spring they would be rounded-up, shod and returned to their district. Sonny Lundwall and I had some semi-trucks and usually had the job of transporting the mules. We hauled mules to Powell, West Fork of the Bitterroot, Bonita, Seeley Lake, Sula and Spotted Bear. Each year the numbers became smaller.

The last year that I had any contact with the Remount and winter range was 1961. It was a long, hot, dry summer and a very bad fire season. Cap Evans called me, saying that he was in a bind and could I come to see him. When I arrived at Cap's office, he asked me to help him locate some pack strings, equipment and packers for the fire season. Fires were breaking out everywhere with smokejumpers and modern equipment unable to stop them. Cap said, "We need men and mules and we need them now." I had a short semi-truck that would haul a full string, plus hay, grain and equipment. It would also go anywhere that an ordinary truck could reach. I would locate a pack string, hire it, deliver it to any place that Cap needed it, and then look for another supply. I delivered strings to the Higgins Ridge fire between Elk Summit and Moose Creek, to the Chicken Ridge fire between Magruder and the Salmon River, to Paradise Guard for the Bear Creek fire, to Rye Creek for the Skalkaho fire, and one string to the West Fork of the Bitterroot. All I hauled for six weeks was pack stock, hay and grain. I hired Red Harper and his string of 10 head of matched pinto horses at Hungry Horse and took them to Elk Summit. Red was packing with Chuck Ray from Whitefish picking up the supplies that were being dropped from the air. They were on a steep hilside trying to turn the strings around when Red got his string tangled in his lead ropes. The fancy pinto horses fell down the hillside, breaking up a lot of stuff and skinning themselves up a lot. Chuck said Red sat down and cried.

A litter for carrying an injured worker in operation.

During the 1961 fire season the Forest Service was hiring pack stock anywhere they could find it and they still could not get enough. While I was scouring the area around here for stock, Ira Cave was sent to Missouri to look for southern mules. He located a couple of carloads and shipped them to the winter range. They were small, narrow little mules that had been used for cultivating corn and cotton. They had never been ridden or packed. The Forest Service saddles were too big for them and required a lot of extra blankets for padding. I went to the winter range to move a string of them to a fire between the West Fork of the Bitterroot and Chicken Ridge. Cap Evans asked me to help the packer get loaded and started to the fire. George Johnson, the packer, was new and inexperienced and so were the mules. We loaded the packs on the mules and George started leading the string as I walked alongside trying to keep the mules trailing one behind the other. The little mules had always been worked in teams and insisted on walking side by side. It took us quite a while to get them going but finally George left for the fire. George told me later that he skinned up a few heads around trees before he got them to string out correctly.

After the fire season was over, I was in Cap Evans office squaring our accounts when Cap made this remark, "We got rid of the mules, and now we know for sure that it was a mistake. I don't care how many machines men

Les Wolf, the last Superintendent of the winter range.

build, neither the Army nor the Forest Service will ever be able to not have some form of animal transportation." The Forest Service has now returned to some pack strings and the Army has re-activated some pack artillery and mountain infantry using dog teams, mules and ski troops.

Some of the leases at the winter range were in effect until 1962. Bob Llewelyn bought the Gustine 1,280 acres; George Stebbins bought the 4,500 acres that was known as the Ferry Basin. Stebbins bulldozed a road down the north side of the Flathead River from the Dixon Bridge to the old corrals and shoeing sheds. This eliminated the use of a ferry. The 3,000 acres in the Pete McDonald place went back to the Flathead tribe; the Mahonet land of about 20,000 acres (the winter range headquarters) was leased to a group of local cattlemen. With the closing of the winter range, the horse and mule operation of the Forest Service became history.

When Cap and I were in his office settling our accounts for the 1961 fire season, we spent more time reliving the history of the Remount and winter range than we did on our bookkeeping. It was an era long gone, never to return.

Packstring information from Northern Region News.

Dixie, Ruby, Festus, Dodge and five other special employees are information ambassadors for the Northern Region, featured at parades, fairs, and other outdoor events. Selection criteria for these positions are not usually considered for Forest Service jobs—age, shape, color, disposition, and how well they entered the "green fleet." Their supervisor promises to keep them in sharp dress, groomed, well-fed, and they will have an opportunity to "see the Region" and other parts of the West. They are the "Region One Pack/Show String" of mules stationed at the Ninemile Ranger District. The string consists of nine matched sorrel mules and a lead horse.

Their role is to demonstrate how traditions endure today in helping manage important backcountry and wilderness areas. "Although their dress is slightly fancier and they are neatly groomed, they will be outfitted for parades with the same type of tack and equipment used in managing our backcountry every day," said Bob Hoverson, Ninemile's packstring manager. "In fact, this team will continue to do regular and special backcountry work wherever needed."

This specially built truck and trailer is used to transport the mules and form the backdrop for a packing demonstration and display area suitable for fairs.

Entrance to the historic Ninemile Remount Depot and Ranger Station.

The barn is the centerpiece of the historic site. It housed animals as well as grain and equipment. Big draft mules lifted hay into the loft using a system of nets and pulleys. Hay was dropped from the loft into a manger. Today the barn is used to store equipment for all aspects of work on the Ninemile District. The Visitor's Center at the historic Ninemile Remount Depot and Ranger Station.

Visitor's Center.

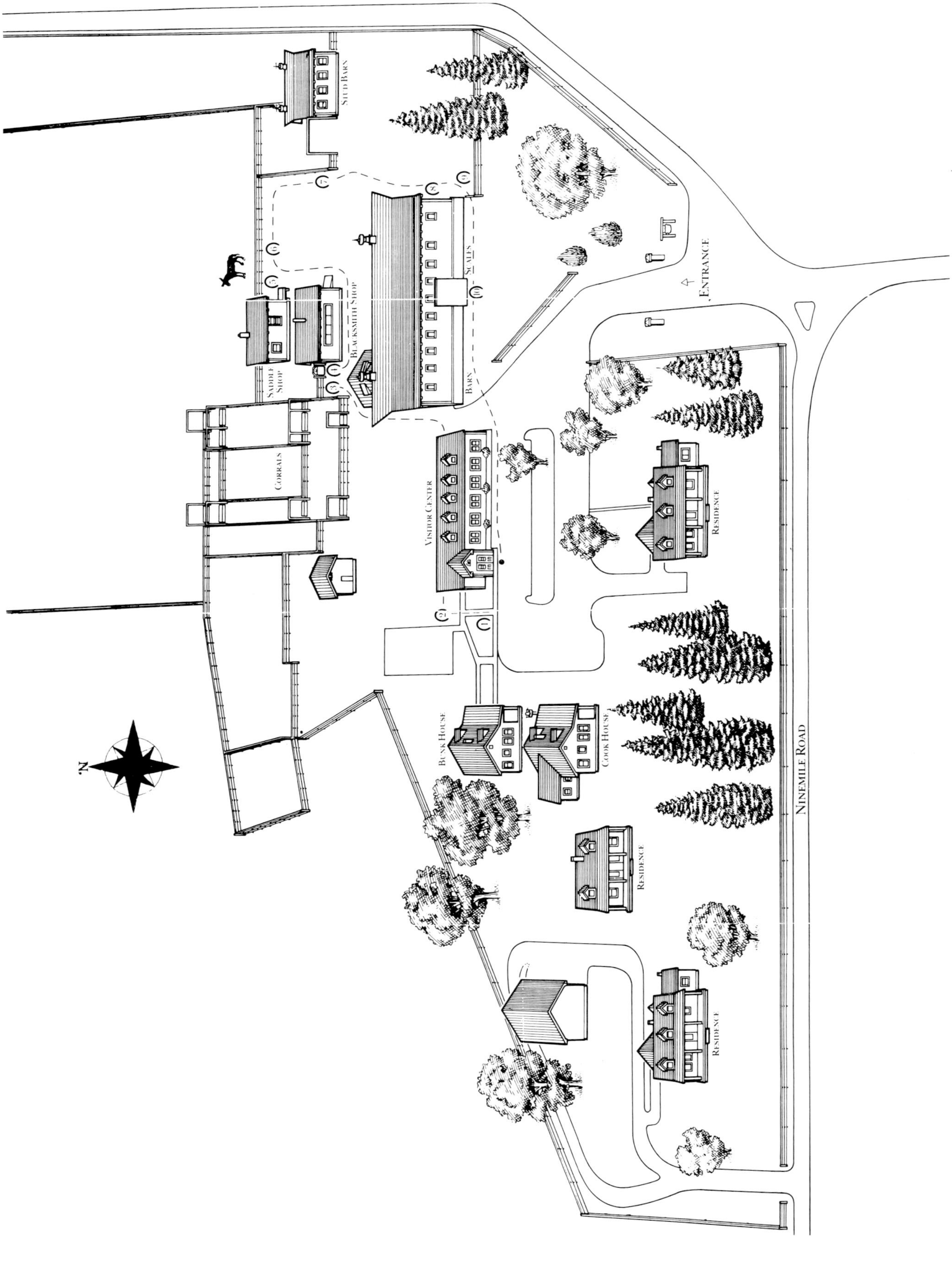
Stud Barn
Scales
Barn
Blacksmith Shop
Saddle Shop
Corrals
Visitor Center
Entrance
Residence
Bunk House
Cook House
Residence
Residence
Ninemile Road
N.

A typical CCC boy in 1939.

CHAPTER FIVE: Civilian Conservation Corps

The Civilian Conservation Corps and the U.S. Forest Service of Region One

Written by Mr. Bill Sharp, ex-Montana CCC enrollee

By April 5, 1933, the U.S. Forest Service Staff of Region One had received orders from the Washington, D.C. office to prepare for an "army" of young men, part of the Civilian Conservation Corps. This relief and welfare work for unemployed men between the ages of 18 and 26 was to take place mainly in the National Forests and Parks. The Federal program was called "Emergency Conservation Work," and the CCC Youth work force was part of this E.C.W. re-employment effort.

Regional Forester Maj. Evan Kelley and his assistant, L.C. Stockdale, headed for Washington, D.C. when they learned that Region One would be able to use CCC labor. By early May some campsites had already been selected and the Army had contracted with local carpenters to set up the buildings. The CCC Company which was to occupy the camp usually sent a construction cadre with a junior officer to furnish the construction labor. The USFS recruited a force of "local experienced men" (LEM) to help with technical problems and to help train the CCC enrollees, and most became "straw bosses" and assistants to the Forestry Foremen Staff. These were older men; many were war veterans, and they were paid $45 per month.

The LEMs became part of the CCC Company and lived with the enrollees. They used their civilian work clothes, but often begged other clothes off the Company Supply Sergeant. Many LEMs were truck and tractor drivers, and some became the camp maintenance men, often looking after the water supply and sewage system. Others were transferred into the Army Company staff as leaders or cooks.

During the mobilization of work project plans for the CCC, Region One enjoyed the convenience of Fort Missoula. The Army activated the Montana CCC Companies and received the companies sent here from the Eastern states. This Fort had an active regular 4th Army Infantry Battalion, and during the summer of 1933 regular Army officers administered the CCC companies. The Montana CCC companies were only part of the larger Ninth Army Corps Area with headquarters at the Presidio in San Francisco. All of the CCC companies of this area (Calif., Ore., Wash., Idaho and Mont.) had the digit nine as the first or second number in their identification. Region One also used CCC companies that were made up and sent out from Fort George Wright in Spokane, Wash.

Region One selected as its first CCC campsite one in the Beaverhead National Forest. It was designated Camp F-1, (forestry camp), and was located on the upper Ruby River at the Vigilante Ranger Station. Montana CCC Co. 955 occupied this camp but only during for the summer of 1933. The men used Army tents over frames with wooden floors, and a mess hall, washroom and latrine were framed up by local carpenters. Temporary tent camps were the rule for 1933, as appropriations for the CCC program did not include long-term work projects. The Beaverhead was also the site for Camp F-2 on the upper Wise River.

An even larger and more important task fell to Region One, and that was the shaping of the work project technical leadership consisting of a superintendent and his forestry foreman. For the of 1933, in most CCC camps, the local

District Ranger became the Project Superintendent. The staff of foremen was recruited by Region One at the forest Supervisor's office. These men were usually veterans of the summer work force. Many were local loggers and sawmill workers or local ranchers and some were from the Butte, Mont. mines. Most were family men and also veterans of World War I. For the most part, they proved very capable in getting the CCC boys to work and work safely, and many were advanced to the Camp Superintendent position by the fall and winter of 1933. This Technical Staff added up to a fair-sized army of professionals who were sent into the woods to 23 Montana CCC camps and many more in Idaho and eastern Washington.

Providing hand tools for CCC boys on the work projects became a big business, and lots of trucks were needed to get the men to work areas. A Ford dealer in Seattle received an order for 82 stake body trucks of which 41 trucks went to Missoula and the rest to camps in the Fort George Wright CCC District at Spokane. The Region also planned many Spike Camps in order to keep the workers close to the project, especially the building of new lookout stations, and since the USFS was responsible for boarding the CCC enrollees in the camps, the Spokane Warehouse had to be well-stocked. For the most part this type of work, away from the main campsite, was treated in the same manner as a fire fighting camp.

The CCC was established in March 1933, to provide employment for some of the millions of youths who could not find work during the depression years. It disbanded nine years later, on June 30, 1942. Over three million men were enrolled in the program through the years.

Fort Missoula CCC Headquarters

This building was built in 1936 as the CCC District warehouse. The foundation for the administration building is to the right. These buildings were later used by the Montana National Guard and are now abandoned. An old 1912 brick commissary building in front of these was used as a CCC warehouse in the 1930s and now houses the Historical Museum at Fort Missoula.
COURTESY DONALD HOBART, SACRAMENTO, CALIF.

An overall view of Fort Missoula in the 1930s. The fort was originally established in 1877 in response to the Nez Perce Indian War. It was occupied off and on through the years by various units of the U.S. Army and was activated in 1933 to be a district headquarters for the CCC camps in most of Montana, Glacier National Park and the winter camps in Yellowstone National Park.
COURTESY MARTIN HUGGINS, KALISPELL, MONT.

Another view of Fort Missoula showing the recreational hall in the center and officer's quarters in the background. The fort site has been divided up amongst several governmental agencies and parts have been sold off for private development. The U.S. Army Reserve, U.S. Navy Reserve and Montana National Guard units still maintain facilities here along with the Historical Museum. RO

Barracks buildings built in 1910. The large building on the left now houses the U.S. Army Reserve unit, the building to the right now houses the Lolo National Forest Supervisor's Office and the Missoula Ranger District. RO

Martin Huggins (left) and Ward Penkake by the old cannon at Fort Missoula, both CCC enrollees at the fort, 1934. At this time enrollees were still using World War I Army issue clothing.
COURTESY MARTIN HUGGINS, KALISPELL, MONT.

The latrine area had 24 water faucets with showers on the south wall. Each man was required to clean up and be ready to go to work early in the morning, five days a week. COURTESY MARTIN HUGGINS, KALISPELL, MONT.

One of the fire engines at the Fort Missoula headquarters.

CCC drivers who worked out of the Fort Missoula garage.

THE GREEN GUIDON June 28, 1935

Fort Missoula, Mont.

☆ 1963 ☆

(Company 1963, Camp Squaw Creek, Bozeman, Montana, June 28, 1935) The new concrete bridge over the west Gallatin river is well on the way to completion. The entire abutment and half the approach on the east side have been finished. On the west side the abutment is half done. The forms are being erected, and the concrete will be poured sometime next week.

We experienced a setback two weeks ago when high water took out the foot bridge. This made it necessary to haul men and materials around by way of the Squaw Creek ranger station, a distance of three miles.

When the bridge is finished it will consist of two approach spans of 26 feet each and a center span of 95 feet 8 inches. The length of the entire bridge will be 147 feet and 8 inches. It will have a width of 18 feet of roadway clearance. At high water the deck will be about 16 feet above the river.

Mr. Henry Teitjen, the engineer, estimates that it will require 520 cubic yards of concrete. This means that 430 cubic yards of crushed rock, 230 cubic yards of sand, and 3120 sacks of cement will be needed. Since the entire bridge, including decking and railing, is to be reinforced, 34 tons of steel will go into it.

The rock crusher and six dump trucks have been working double shift for the past two months. All of the rock and sand necessary to complete the job has been hauled and placed in stock piles near the abutments.

The concrete now poured is tested up to 3,500 pounds per square inch. The bridge, which is being built to carry a 30 ton load, is to be finished and ready for traffic on the 25th of next September.

THEY'RE IN THE ARMY NOW

Montana Standard
May 2, 1933

First Montana C.C.C.'s for Fort Missoula ---

BUTTE RECRUITS OFF TO "FRONT LINES."

Grinning over their forthcoming experiences in the emergency conservation corps these Butte lads are shown as their train rolled out of the Northern Pacific station yesterday for Missoula where they will take preliminary training at Fort Missoula before being sent into the forests. There were 106 youths in Butte's quota, which was examined yesterday at Legion hall by Capt. J. D. Pomerene of the Fourth infantry who was assisted by Col. D. G. Stivers, representing the Silver Bow County Emergency Relief association. All of the recruits are between 18 and 25 years of age.

When the CCCs arrived in Region One, there were 17 National Forests under the Missoula office. Twelve of these were in Montana, with five in the Idaho panhandle and eastern Washington. All but two of these forests were selected for CCC camps during the summer of 1933. These were the Helena and the Absarokee National Forests. However, as the work program of the CCC was extended through the 1930s, all of the National Forests did gain a camp, at least for the summer season.

Region One had 12 fire fatalities in its CCC companies. Eight of these fire losses were in the Panhandle section of Idaho. Seven of the eight losses were from falling snags on the fire line and one was a heart failure of a CCC-LEM on the fire line.

Five of Region One's National Forests were in the Idaho Panhandle—Nez Perce, Clearwater, Coeur d'Alene, Kaniksu and the St. Joe Forests. Through the CCC era they accumulated some 80-plus CCC campsites, which were under two CCC Districts. In these forests the CCCs have been given credit for building the first roads along the St. Joe, Clearwater, Selway and Lochsa rivers. It was true that the CCC roads were mainly "truck trails," but a good share of the work was manual labor, even drilling blasting holes with hand drills, digging in culverts, building bridges, backslopping and building retaining walls. The CCCs from Camp F-204, Falls Creek at Hess, Idaho, built a long steel bridge, spanning the St. Joe River, the Milwaukee Railroad and a forest road. White pine areas of the Idaho forests had many crews working on Blister Rust Control. In the Priest River old burns, many snags were removed and trees thinned for the fire hazard reduction plan.

Region One entered the 1934 spring season with six winter camps, three on the Cabinet and three companies of the large Ninemile Stoney Creek Camp

WRITTEN BY THE CCC FOR THE CCC

HAPPY DAYS EVERY WEEK

Happy Days

SEND HAPPY DAYS HOME

CORPS WILL BE EXPANDED TO 299,800

CCC Gets More Ratings as Co's. Are Boosted to 200 Men

25,000 EASTERN MEN TO BE SENT WEST

Will Need 100,000 Men to Fill Ranks After June 30

Capt. Leaves CCC After 61 Months of Service

No Three Month's Service in West

First It Was LEM's; Now It'll Be PA's

Harvard Botany Professor Critic of CCC Declines Two Bids to Visit Camps

Over the nine year period of the CCC, more than 4,000 camps were built in the United States on National Forest, National Park, reclamation, TVA Soil Conservation Service, Corps of Engineers, state park and other government lands. This map shows camps as of Aug. 1, 1938.

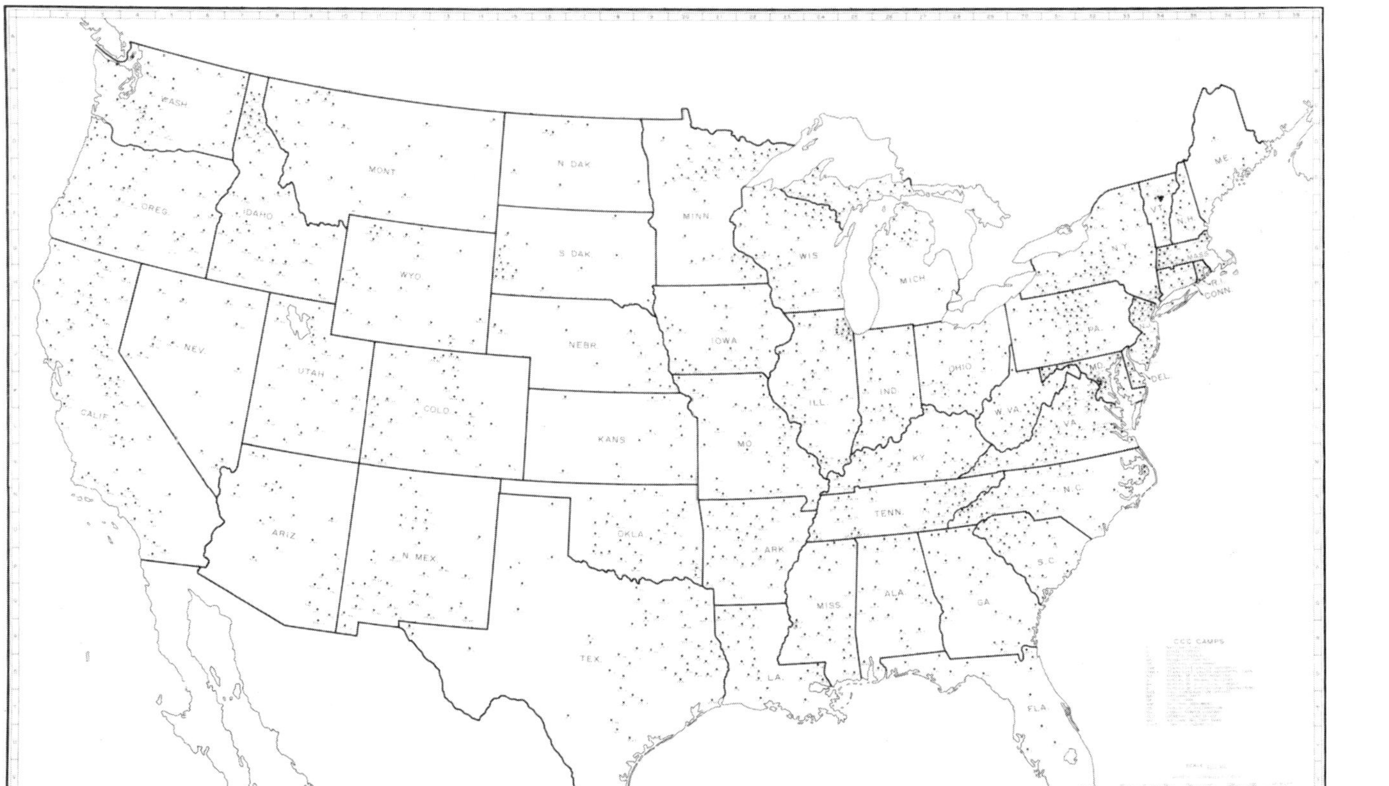

on the Lolo. The 12 National Forests in Montana began formulating plans for a 1934 summer season of CCC workers. As before, most summer camps would be of the tent type, but a few forests began construction of a more permanent barracks-type camp. Ninemile camp was used for a staging area for Montana CCC men returning from California camps. Of the 12 National Forests in Montana and under the Ft. Missoula CCC District, Helena and Absarokee National Forests again were not assigned CCC camps during the second season, and Lewis & Clark and Custer forests lost their active 1933 camps. The Region might have been assigned more Montana CCC companies, but several had to be shared with other agencies and with Glacier National Park. By the spring of 1934 appropriations from Congress extended the CCC for another year as President Roosevelt authorized continuance to March 31, 1935.

Suspension tower at the south end of the Manning Bridge, built by the CCC in 1934, Nezperce National Forest. The bridge was named for a CCC boy who died.

By April 1935, which was the end of a six-month enrollment period for enlistees, Region One was preparing for new CCC recruits and making plans for summer camps. Many of these camps would be of a barracks-type, capable of winter occupation. This was partly due to the effort of Relief Administrators to spend more for material and to hire contract carpenters in order to create more jobs. New camps also made Region One appear to place the CCCs in a permanent place, with the large repair shops in Missoula hiring CCC help. Many Forest Supervisors were planning long term work projects. President Roosevelt wanted a permanent CCC, and forest service planners seemed to agree with him. The Army continued to improve the camp life of the enrollee. There were no more mattresses of wool sacks, stuffed with straw, and no more double bunking in the barracks. The educational program was expanded with nearly all the forestry foremen, superintendents and Army officers as instructors. Athletic events and social functions were part of every evening; there were longer weekend trips. The last CCC camp for Region One, USFS and the Fort Missoula CCC District was built in the late 1940s on the Lolo National Forest. It was numbered F-81 and was meant to replace the old triple CCC Co. Camp Ninemile. On April 13, 1941, after winterizing was completed, CCC Co. 1962, made up of Montana men moved in from the Ninemile Camp. The new camp was less than a mile from the USFS Remount Station and was named "Grand Menard" for one of the station's stallions. The camp remained active until July 31, 1942, the end of the CCC era. Toward the end the camp became the headquarters for the disbanding of CCC companies. CCC crews from this camp continued to do work at the Remount pastures, and a large Spike Camp was part of the winter range near Perma, Mont. As the war effort intensified and Montana's CCC companies disbanded, some of the enrollees with more than 90 days to serve were sent to Camp F-81.

A report out of Washington, D.C. stated that $19,000,000 was spent in Montana through the CCC program and that 20,977 Montana men served in the corps from April of 1933 on.

Company 953, F-47 compliment of enrollees and administrative personnel at Packer Meadows, Lolo N.F., June 6, 1935. UM

Camps

The first camp established in Region One was on the upper Ruby River at the Vigilante Ranger Station on the Beaverhead National Forest (F-1), summer, 1933.

Camp P-210 (private land) at Bovill, Idaho, appears to have been housed in a logging camp.

Bungalow CCC Camp on the Clearwater River, Clearwater National Forest, 1939.
LOCHSA R.D., LLOYD PARKER PHOTO

Camp F-44 near Clarkia, Idaho, on the Cabinet National Forest.
IDAHO PANHANDLE N.F.

A camp at French Creek on the Nezperce National Forest in Idaho. For a kid from the streets of a large eastern city living in this environment would be a complete and challenging change.

The Tank Creek Spike Camp on the Lolo National Forest. Spike camps were temporary camps used for either summer or remote projects away from the main camp. Tents were set up for living quarters.

Nine Mile Camp

An aerial view of the Nine Mile Camp in August 1941. By this time the camp was being used as a training center for smokejumpers and firefighters. RO

Entrance to the Nine Mile Camp, the largest in the United States.

Barracks at Nine Mile. A typical camp had about 200 enrollees, but Nine Mile had over 400 at one time. Each barracks building, constructed of wood frame, could accommodate 40 to 50 men, with heat provided by wood stoves.

Headquarters buildings at Nine Mile. Each company had a company commander, who was either a regular army reserve officer, plus a junior officer, camp doctor and educational advisor. The project superintendent, usually employed by the technical service the camp was under, was in charge of all work projects away from camp and had eight to 10 foremen under him.

It was not all work at the camps. Here some boys are preparing for a "snipe hunt" probably for some new, unsuspecting "boys" at the Nine Mile Camp, 1933.

A fencing crew at work at the remount depot pasture, Lolo National Forest, 1934.

The unfinished buildings to the right are additions to the recreational hall, Nine Mile Camp, 1934.

Above and below: getting ready for the day's work, Company 1341, Nine Mile Camp 934.

Starting in 1934 map-making and drafting were taught at the Nine Mile Camp. The CCC program was initially set up as a make-work program but as things expanded, more educational opportunities were developed for the enrollees and in the latter years, many different courses and skills were being taught.

SUTLIFF IN PRAISE OF TWO CCC YOUTHS

Returns to His Home From Hospital. Johnson Recovering From Crash.

Hamilton, Aug. 25.—Clarence B. Sutliff went to his home from the Daly hospital Thursday and is recovering from injuries received in a plane crash on Roaring Lion hills Tuesday night.

Today he had more words of appreciation for the help given him and Pilot Dick Johnson by the CCC boys, Robert Bloom and William Johannes, truck drivers from the Trapper creek camp, who happened along the trail while off duty, bound for the fire scene in an attempt to snap some night pictures. A camera and flashlights were brought along, but the camera was forgotten as Bloom went back to the base camp for help and Johannes climbed the knife-blazed trail up the mountain to Pilot Johnson, sick and injured too badly to move.

Late word today from the Daly hospital said Johnson was continuing to improve from his injuries—fractured ribs, bruises and a concussion.

Supervisor G. M. Brandberg said at 4 o'clock that the fire was under control and while it burned spottedly over a 4,000-acre area, that actually from 800 to 1,000 acres had been affected. The entire creek bottom was cleaned off by the fire, he said. The supervisor telephoned the information from the base fire camp in the east end of the canyon.

Tree planted in memory of Robert Fechner by CCC corpsmen at the junction of Highways 3 and 10 near Missoula, Mont.

Making fence bucks for the remount depot pasture lands, March 1934.

Wood work along the Clark Fork River near Nine Mile, March 1934.

Digging an irrigation ditch for the remount depot hay pastures.

All that is left of an officer's quarters at the Nine Mile Camp.

The foundations for the barracks at the Nine Mile Camp can still be seen.

In 1934 a sign shop, staffed by CCC enrollees from Fort Missoula, was established in downtown Missoula at the Park Hotel, facing the Northern Pacific Depot.
COURTESY MARTIN HUGGINS, KALISPELL, MONT.

Camp Life and Work

F-2, Wise River Camp on Little Joe Creek, Beaverhead National Forest, administrative personnel, 1933. From left: Capt. Halbert, Lt. Sabin, Earl Sandrig, Capt. Bedell, Capt. Barker. MHS

Road construction and telephone line crew from Wise River, 1933. Most of these "boys" were high school enrollees from the Butte area along with 50 LEMs. MHS

F-27, Squaw Creek Camp, Gallatin National Forest, crew getting instructions, Jan. 9, 1935. This camp was located in the Gallatin Canyon.

Elk River CCC planting crew, 1936.
LOCHSA R.D., LLOYD PARKER PHOTO

Building a fence at the extension project, Savenac Nursery, Cabinet National Forest at Haugen, Montana.

Technical instruction was a major part of the CCC program in the latter years of 1939-42. Many of the enrollees used their new skills throughout their working years.

Enrollees from the Bernice Camp (F-13) near Basin, Montana, Deerlodge National Forest, hand-drilling blasting holes, September 1933.

Drilling blasting holes on the Paradise-St. Regis Road project, Cabinet National Forest, March 1934.

Dumping rock on the Paradise-St. Regis Road project, Cabinet National Forest, March 1934.

Workers on the Salmon Mountain Road, Bitterroot National Forest, July 1933.

By the summer of 1940, President Roosevelt proclaimed a limited national emergency after Germany overran most of Western Europe and as a result the CCC became more involved in national defense work. The program was no longer just a conservation work force with courses being instituted to provide training in skills needed in defense or military work. These photos, all taken in the Missoula area, show training in operating a Forest Service SPF radio set (an intensive six-month course) and an auto mechanics course (12 months). After the Pearl Harbor attack, the CCC offered all its camps to the Army. Except for forest fire fighting, all conservation work off military reservations was cancelled on Jan. 1, 1942. One hundred camps were assigned to fire duty in the West but the manpower drain was tremendous as thousands of enrollees and administrators left to enter military service. No funds were appropriated for the CCC after June 30, 1942.

CCC enrollees participate in a parade in Coeur d'Alene, Idaho in the late 1930s.

EX — C.C.C.

My feet are on the pavements once again,
  But though the same old search begins anew,
I cannot shake the wonder from my brain,
  —Remembering the things that I've been through.

My days are cheered by bouyant memories;
  The ring of ax and pick still haunt my ears.
I see the road we built between the trees
  (The first great effort of our youthful years.)

My feet are on the pavement once again,
  But in my heart a newer courage stirs,
A hope—a trust—, Can anyone explain
  Just what I found out there among the firs?

John Urbanek,
Camp F-10, C.C.C.,
Trout Creek, Montana

The *Green Guidon* issued by the CCC at Fort Missoula in the late 1930s.

Camp Birch Creek in the Beaverhead National Forest in 1935. The camp was near Dillon, Montana, and is still in use under a special use permit to Western Montana College and is used for outdoor education.

Camp Needmore near Ekalaka, Montana, in the Custer National Forest in 1935. The camp is still in use under a special use permit to the Custer County Commissioners. It is reserved for 4-H clubs and Ekalaka community use. MONTANA HISTORICAL SOCIETY

0
CCC
Camp Program

CIVILIAN CONSERVATION CORPS CAMPS IN REGION ONE
Tenth Enrollment Period (October 1, 1937 to March 31, 1938)

November 1, 1937.

Camp Number	Company Number	Camp Name	Forest	Army Dist.	Mail Address	Railhead	Camp Superintendent
N. Idaho							
F-123	5702	North Fork	Clearwater	L.	Pierce, Idaho	Jaype, Idaho	R. L. Valentine
F-193	5705	Bungalow	"	L.	Pierce, Idaho	Jaype, Idaho	L. J. Morin
P-276	5703	Brown's Creek	State (Clw.)	L.	Pierce, Idaho	Greer, Idaho	F. W. Pearce
S-279	5706	Dent	" "	L.	Ahsahka, Idaho	Ahsahka, Idaho	O. C. Hopkins
F-132	531	Big Creek #2	Coeur d'Alene	G.W.	Prichard, Idaho	Enaville, Idaho	G. B. Valentine
F-154	562	Devil's Elbow	"	G.W.	Prichard, Idaho	Enaville, Idaho	Leonard Morris
F-180	3266	Hayden Creek	"	G.W.	Coeur d'Alene, Ida.	Coeur d'Alene, Ida.	D. M. Lightner
F-182	1249	Wolf Lodge	"	G.W.	Coeur d'Alene, Ida.	Coeur d'Alene, Ida.	Wayne Newcomb
F-107	1293	Pollock	Nezperce	L.	Riggins, Idaho	Grangeville, Idaho	A. W. Foskette
F-115	1224	Slate Creek	"	L.	Whitebird, Idaho	Grangeville, Idaho	Blaine Snyder
F-190	5704	O'Hara	"	L.	Lowell, Idaho	Kooskia, Idaho	Delbert Cox
F-117	257	Marble Creek	St. Joe	G.W.	Calder, Idaho	Marble Creek, Idaho	Ben Bates
F-120	1204	Avery	"	G.W.	Avery, Idaho	Avery, Idaho	C. A. Gregory
F-187	1239	St. Joe River	"	G.W.	Avery, Idaho	Avery, Idaho	E. D. Neiland
F-188	229	Willow Creek	"	G.W.	Emida, Idaho	Santa, Idaho	Andy Porterfield
F-203	4762	Drysdale	"	G.W.	Calder, Idaho	Herrick, Idaho	Herb Richey
F-127	1235	Experiment Sta.	Kaniksu	G.W.	Priest River, Idaho	Priest River, Idaho	Louis H. Whetsler
F-142	4798	Kalispell Bay	"	G.W.	Priest River, Idaho	Priest River, Idaho	G. H. Parker
F-159	1205	Blowdown #2	"	G.W.	Priest River, Idaho	Priest River, Idaho	A. S. McQuilken
F-164	594	Four Corners	"	G.W.	Priest River, Idaho	Priest River, Idaho	Maurice Kelley
E. Washington							
F-1	2920	Sullivan Lake	Kaniksu	G.W.	Metaline Falls, Wn.	Metaline Falls, Wn.	P. J. Sullivan
F-102	281	Kalispell Creek	"	G.W.	Priest River, Idaho	Priest River, Idaho	H. B. Craig
Montana							
F-60	1501	Birch Creek	Beaverhead	M.	Dillon, Mont.	Apex, Mont.	W. F. Edwards
F-40	1961	St. Regis	Cabinet	M.	St. Regis, Mont.	St. Regis, Mont.	W. V. Jefferson
F-52	1998	Thompson River	"	M.	Thompson Falls, M.	Thompson Falls, M.	H. F. Johnson
F-75	4782	Flint Creek	Deerlodge	M.	Phillipsburg, Mont.	Phillipsburg, Mont.	L. L. Scharmota
F-57	1963	Squaw Creek	Gallatin	M.	Bozeman, Mont.	Bozeman, Mont.	J. A. Anderson
F-76	530	Libby	Kootenai	G.W.	Libby, Mont.	Libby, Mont.	G. M. Daugharty
F-61	1999	Belt Creek	Lewis & Clark	M.	Neihart, Mont.	Neihart, Mont.	W. F. Krukeberg
F-42	591	Oregon Butte	Lolo	M.	Superior, Mont.	Superior, Mont.	Russell Peterson
F-55	2945	Lolo Creek	"	M.	Lolo Hot Springs, M	Missoula, Mont.	Wm. Roullier
F-9	956	Taft	Nursery	M.	Haugan, Mont.	Haugan, Mont.	Paul Nelson
F-36	1962	Ninemile #1	Remount	M.	Huson, Mont.	Huson, Mont.	L. Vierhus
F-37	952	Ninemile #2	"	M.	Huson, Mont.	Huson, Mont.	L. Vierhus
F-38	939	Ninemile #3	"	M.	Huson, Mont.	Huson, Mont.	L. Vierhus

Key to Army Districts: L. - Lewiston; G.W.- Fort George Wright; M. - Fort Missoula.

Idaho F Camps	18	
Idaho S & P Camps	2	20
Washington F Camps	2	2
Montana F Camps	13	13
Total Camps R-1		35

CCC
ORGANIZATION
Directories

June 1, 1939.

CIVILIAN CONSERVATION CORPS CAMPS IN REGION ONE
Thirteenth Enrollment Period (April 1, 1939 to September 30, 1939.)

Camp Number	Company Number	Camp Name	Unit	Army Dist.	Mail Address	Railhead	Project Superintendent
N. Idaho							
F-123	5702	North Fork	Clearwater	G.W.	Pierce, Idaho	Jaype, Idaho	Blaine Snyder
F-193	5705	Bungalow	"	G.W.	" "	" "	Lee J. Morin
F-132	531	Big Creek	Coeur d'Alene	G.W.	Prichard, Idaho	Enaville, Idaho	G. B. Valentine
F-137	1249	Deception Creek	"	G.W.	Enaville, Idaho	Coeur d'Alene, Ida.	W. S. Newcomb
F-154	562	Devils Elbow	"	G.W.	Prichard, Idaho	Enaville, Idaho	L. J. Morris
S-263	281	Priest Lake	Idaho State	G.W.	Coolin, Idaho	Priest River, Idaho	Oscar C. Hopkins
P-276	5703	Brown's Creek	" "	G.W.	Pierce, Idaho	Greer, Idaho	Frank W. Pearce
F-190	5704	O'Hara	Nezperce	G.W.	Lowell, Idaho	Kooskia, Idaho	Delbert Cox
F-192	3275	Red River	"	G.W.	Elk City, Idaho	Grangeville, Idaho	A. W. Foskette
F-187	1239	St. Joe River	St. Joe	G.W.	Avery, Idaho	Avery, Idaho	Geo. David
F-188	229	Willow Creek	" "	G.W.	Emida, Idaho	Santa, Idaho	A. C. Porterfield
F-203	3272	Drysdale	" "	G.W.	Calder, Idaho	Herrick, Idaho	H. J. Richey
F-204	1204	Falls Creek (Hess)	" "	G.W.	St. Joe, Idaho	St. Joe, Idaho	C. A. Gregory
F-205	257	Ruby Creek	" "	G.W.	Bovill, Idaho	Bovill, Idaho	Ben Bates
F-127	1235	Experiment Station	Kaniksu	G.W.	Priest River, Idaho	Priest River, Idaho	L. H. Whetsler
F-159	3277	Blowdown	"	G.W.	" " "	" " "	G. H. Parker
F-164	594	Four Corners	"	G.W.	" " "	" " "	Maurice Kelley
E. Washington							
F-1	2920	Sullivan Lake	Kaniksu	G.W.	Metaline Falls, Wn.	Metaline Falls, Wn.	Donald O. Sands
F-103	2936	Usk	"	G.W.	Usk, Wn.	Usk, Wn.	P. J. Sullivan
Montana							
F-60	3282	Birch Creek	Beaverhead	M.	Dillon, Montana	Apex, Montana	W. F. Edwards
F-78	1501	Trapper Creek	Bitterroot	M.	Darby, Montana	Darby, Montana	W. A. Rooney
F-52	1998	Thompson River	Cabinet	M.	Thompson Falls, M.	Thompson Falls, M.	H. F. Johnson
F-80	297	Palisades	Custer	M.	Red Lodge, Mont.	Red Lodge, Mont.	A. G. Tucker
AI-1	2749	Spotted Eagle	Custer**	7th C.A.	Miles City, Mont.	Miles City, Mont.	Breeze Boyack*
F-75	3281	Flint Creek	Deerlodge	M.	Philipsburg, Mont.	Philipsburg, Mont.	L. L. Scharmota
F-77	952	Columbia Falls	Flathead	M.	Columbia Falls, M.	Columbia Falls, M.	W. V. Jefferson
F-57	1586	Squaw Creek	Gallatin	M.	Bozeman, Montana	Bozeman, Montana	James A. Anderson
F-79	3266	Rimini	Helena	M.	Helena, Montana	Helena, Montana	Wm. O. Roullier
F-76	530	Libby	Kootenai	G.W.	Libby, Montana	Libby, Montana	G. M. Daughtery
F-61	1999	Belt Creek	Lewis & Clark	M.	Neihart, Montana	Neihart, Montana	Henry P. O'Donnell*
F-42	591	Oregon Butte	Lolo	M.	Superior, Montana	Superior, Montana	R. H. Peterson
S-208	3271	Stryker	Montana State	M.	Stryker, Montana	Stryker, Montana	James Lee
F-9	956	Taft	Nursery	M.	Haugan, Montana	Haugan, Montana	Paul Nelson
F-36	1962	Ninemile	Remount	M.	Huson, Montana	Missoula, Montana	L. M. Vierhus

*Effective 8-1-39.
**Bureau of Animal Industry.

Key to Army Districts: G.W. - Fort George Wright. M. - Fort Missoula.

Idaho F Camps	15	
Idaho S and P Camps	2	17
Washington F Camps	2	2
Montana F Camps	13	
Montana AI Camps	1	
Montana S Camps	1	15
Total Camps R-1		34

Civilian Conservation Corps camps in Montana and under Fort Missoula CCC District

April 1933 to July 1942

Address	Camp Name	Project	Camp Number
West Yellowstone	Basin R.S.	Gallatin N.F.	F 16
West Yellowstone	Basin R.S.	Gallatin N.F.	DF 73
White Sulphur Sp.	Duck Creek	Helena N.F.	F 59
Winnett	Winnett	Soil Conservation Service	SCS #1
Winifred	Winifred	Department of Grazing	DG 71
Whitehall	Caverns (LaHood Park)	State Parks	SP #3
Whitehall	Whitehall	Department of Grazing	DG 127
Wise River	Wise River (Little Joe Cr.)	Beaverhead N.F.	F 2
Whitefish	Talley Lake	Flathead N.F.	F 5

CCC camps in Glacier National Park - Administered by Fort Missoula CCC District

G.N.P. Camp No.	Camp Name	Post Office
NP #1	McDonald Creek	Belton, Montana
NP #2	Apgar	Belton, Montana
NP #3	Fish Creek	Belton, Montana
NP #4	Many Glacier	Babb, Montana
NP #5	No Name Creek	Babb, Montana
NP #6	Two Medicine	East Glacier, Montana
NP #7 (no details)		
NP #8	Anaconda Creek	Belton, Montana
NP #9	Belton (Hdq. Co.)	Belton, Montana
NP #10 (no details)		
NP #11	St. Mary's Lake (Roe's Cr.)	St. Mary, Ronan
NP #12 (no details)		
NP #13	Sherburne Lake	Babb, Montana
NP #14	Round Prairie	Polebridge, Montana
NP #15	Apgar Flats	Belton, Montana
NP #16 (no details)		

CCC Camps in Yellowstone National Park - Administered by Fort Missoula CCC District
(During Winter Months - October through May)
(Summer Season camps under Pocatello, Idaho, CCC District)

Y.N.P. #1	Mammoth	Y.N.P., Wyoming
Y.N.P. #7	Glenn Creek	Y.N.P., Wyoming

Three CCC Districts shared Montana camps:
Eastern Montana CCC camps - Ft. Lincoln, Bismarck, North Dakota
N.W. Montana CCC camps - Ft. Geo. Wright, Spokane, Washington and Camp Crook, So. Dakota, Camp F-19, Custer N.F. on State Line - Ft. Meade, So. Dakota.

CHAPTER SIX:

Range

Sappington and Harrington herd of registered cattle in a pasture near the Jefferson Rover in 1943, near Sappington, Montana.

Range

In the late 1870s and early 1880s there were still thousands of buffalo being slaughtered in Montana. By the end of the 1880s, however, they were gone. As fast as they were slaughtered, they were being replaced. The buffalo served as livestock for the Indian, but the new European settlers wanted to see cattle, sheep and horses using the range.

The Indians were confined to reservations soon after their battle with Custer in 1876. By 1878 the Army felt the area that now includes the Northern Region was safe for homesteading. Soon thousands of cattle were driven into the area from Texas, followed by hundreds of thousands more. The grass was excellent. The land was available to anyone strong enough to use it and keep others out. Life on the plains was good. All you had to do was turn cows loose on the range, round them up in the fall, and sell everything you did not want to turn back on the range.

Rustlers and the severe winter of 1886-87 changed the concept of only open-range. Because cattle losses from open ranges were very heavy, cattle raising became more controlled. Hay was put up and fed in winter. Rustlers were jailed, hung or run off. Sound livestock and land management systems began to take shape.

Before the turn of the century sheep began arriving in large numbers. Wars between the sheepmen and cattlemen erupted. This fueled the pens of countless writers who described this epic struggle in prose, poems and song. There have probably been more cowboys and sheepherders killed in stories of this period than ever actually existed! It was an exciting time, and the stories had a foundation in fact. Many of the conflicts occurred about the time the Forest Reserves were created and became National Forests.

For example, in 1901 the January 3 *Yellowstone Journal* reported 11 masked men raided the Daut and Selway sheep camp. The raiders clubbed about 2,000 sheep to death before riding off. On January 10, the *Yellowstone Journal* reported the sheriff was totally unsuccessful in identifying any of the raiders. Raids of this nature were not uncommon in those days. This was the time when many powerful stockgrower associations, groups that excluded sheepmen, were formed to preserve the Western Ranges for cattle.

Albert E. Cole, who was a Ranger on the Deerlodge National Forest, tells a story about the cattle vs. sheep situation that had an ironic twist:

> "One incident had occurred in the summer of 1904 when one outfit that ran both classes of stock had one of their sheep outfits raided, the herder beaten and told to clear out, their sheep scattered through the timber, the camp burned and everything destroyed by cowboys belonging to their own ranch. The herder made his way back to the home ranch, but not before the cowboys had returned and boasted of their exploit. About the first man the herder met when he got back to the home ranch was the foreman, and as the herder had never been to the home ranch before, having been outfitted on a small unit of the company ranches, the foreman asked him very profanely what he was doing there. The herder told his story of the raid and said he had been herding for this outfit. The foreman and his cowboys were wearing large bandanna handkerchiefs over their faces when they raided the sheep camp, so the herder did not recognize the foreman. The foreman recognized the herder all right but did not let on to him who had raided him. The incident was kept quiet for a long time, but the facts finally

leaked out and other sheep outfits used the story as propaganda against the bloodthirstiness of cowboys in general."

In 1905, Gifford Pinchot was aware that many western ranchers and stock growers were more than just concerned about the creation of the National Forests—they were flat out-and-out opposed to them. Many of them had been using these public lands for years—and without interference from the government. The United States government was retaining ownership of summer range lands by creating Forest Reserves and National Forests.

Many ranchers in the West needed these lands for their operations, and they considered the government dictating how the lands should be managed to be real threat.

According to Albert Cole:

> "Almost all the ranchers of that part of the country (Ruby Valley in Montana) were very hostile to the Forest Service objectives and were very bitter in their denunciation of the rules, but were not necessarily hostile or dangerous to us as individuals nor to us as Forest Service employees, as they knew we had to do our job and they realized we had no personal feeling toward them in the performance of our duties. They all thought that the Forest Reserve idea was a crack-pot scheme of some politicians in Washington and that it would be done away with as soon as a Democratic administration took over the government of the country, and even President Theodore Roosevelt, who had been quite popular in that country before, was severely criticized and lost considerable prestige and support because he supported the movement."

Rangelands have always been an important part of the National Forests in the Northern Region. Ranchers grazing livestock on public lands in the West existed before the first Forest Reserves were created in 1891. Applying principles of conservation and regulating livestock grazing was not well received and was a challenge for early day Forest Service officers.

Managing public range, while posing many problems, gave Gifford Pinchot a chance to establish himself and a chance to prove the merits of conservation. In 1898 the Secretary of the Interior, because little was known about the effects of grazing in forests, prohibited grazing in the Forest Reserves. Most ranchers agreed range studies were needed, but they did not want a moratorium while they were being completed. The Secretary of the Interior then modified the Department's position and granted use of the "accustomed ranges."

Gifford Pinchot, then Chief of the Division of Forestry in the Department of the Interior, insisted that his field agents be impartial, because, after all, "there were no preconceived ideas to establish or to defend." He felt that forage should be harvested by grazing and not be wasted, and that grazing should be regulated, not prohibited. He wanted all questions about grazing to be decided on their merits.

Pinchot then informed his agents that there were radical differences of honest opinion they would have to contend with, but he later learned that sometimes the stockmen's opinions were not all that honest. One of his agents was offered a $10,000 bribe to report that sheep grazing was detrimental to the range.

Pinchot worked hard and effectively to explain conservation to representatives of the livestock industry and gained the respect of their leaders. Many of the bonds developed between the stock grower associations and the Forest Service during these formative years are just as strong today as they were when they were forged 80 to 90 years ago.

As an example, Northern Region retiree John Lowell provides an interesting account of Pinchot's attendance at a Western Slope Stock Growers Convention in Glenwood Springs, Colorado, in 1906:

"There were a lot of speeches and resolutions attacking the (Forest) Service and its officers, and it looked like the meeting would close with still further antagonism to the Forest administration. I shall never forget the quiet, convincing attitude and speech of Mr. Pinchot. He was a tall, lean, striking figure at that time. His cool logic, complete information with not undue persuasiveness, and evident determination to stand by the policies and regulations of the Service, brought about a tremendous change. While all the stockmen had regarded him as their arch enemy, before the meeting closed many, if not a majority of them, changed their attitudes entirely. Some, because they realized they would have to comply; and some because they were beginning to believe that after all reasonable regulation of stock grazing on the Forests would in the end be to their interest."

In 1906, stock growers initiated test cases in the courts hoping to end the Forest Service regulation of their use of forage in the National Forests. One case, U.S. v. Light involved cattle in trespass and the other, U.S. v. Grimaud, involved sheep in trespass. In 1911 both cases were settled by the Supreme Court in favor of the Forest Service.

Pinchot's 1907 publication, "The Use Of The National Forests," covers range management:

Early ranchers conquered grizzly bears, wolves, bad weather and rustlers, while assuming a right to graze their livestock on rangelands in the public domain. Many considered the early Forest Service and conservation another battle they had to win.

TO THE USER OF THE RANGE

What happens to the range? Most of the timber land in the West is good for livestock. This range has to be included in the National Forests, because it goes with the timber and cannot be separated from it.

Is it shut out from use? Quite the contrary. It is grazed by cattle, sheep, and horses just as it always has been. It is one of the resources and is there to be used. At present it is used by about 1,500,000 cattle and 6,000,000 sheep. The government protects it from being burned up or from being overcrowded and overgrazed, prevents disputes between the owners of stock, and sees each owner gets use of that range to which he has the best right. The small man with a home in the National Forest always gets the first chance.

In another section, the publication goes into greater detail describing management of range in the National Forests:

TO KEEP THE RANGE IN GOOD CONDITION

The use of the range by live-stock enters unavoidably into the management of the National Forests. All through the western mountains the range goes with the timber; it cannot be separated from it. It is a great resource, and of course ought to be used. The way it is used has a great deal to do with the growth of young timber and the flow of water. It is not wasted or used up, but wisely used, it neither harms the forest growth nor has it a bad effect on the water flow. If it is overgrazed or destroyed, the young growth is stamped down or eaten off, and the soil is left unprotected, to be washed down the slopes into the canals and reservoirs below.

In the use of the range the National Forests work first to protect the settler and home builder. They make sure, before everything else, that he has what range he needs for his own stock. Before the Forests were made he was at the mercy of the big stockman who often drove his herds in from a distance and completely grazed off the settlers range right at his own door. This cannot happen in a National Forest, because the man with a home is sure of the range nearby for his own use, and the big men from a distance are kept away from him.

In allocating the range the small local owners are considered first; then the larger local owners who have regularly used it; then the owners who live at a distance, but who have been regular occupants; and lastly, if there is any room left after these have been provided for, the owners of transient stock.

Special effort is being made to keep down wild animals which damage stock, and the Forest officers, when requested, help to enforce the State and Territorial livestock laws.

A small fee is charged for grazing on the National Forests because when any man gets for his own special use any property maintained for the use of the whole people, he ought to pay for it. Most people are quite willing to pay the cost of restoring the range and keeping it in good condition, especially when such control does away with the old conflicts of all kinds and assures each man of getting his rights. The men who use the range are not the kind who think they ought to get something for nothing.

National Forests, then, are not made for the special purpose of controlling the live-stock business; they are concerned with it incidentally, and help to regulate the use of the range because the people want it regulated.

Finally, "The Use Of The National Forests" explains how the grazing permit system works:

THE USE OF THE RANGE

To secure a permit to graze live-stock on a National Forest, apply to the Supervisor. The nearest Ranger will furnish an application blank. The Supervisor grants the permit. Every year before the grazing season opens, notice is given of the date on which all applications must be in. Be sure to send your application in on time. The Supervisor or the Rangers will furnish all necessary information about the fees and other details of the business.

No stock can be grazed without a permit, except the few head in actual use by campers, prospectors, and travelers; ten head of milch cows or work animals owned by settlers living in or near a Forest, and a reasonable number of saddle, pack, and work animals used for caring for stock grazed under permit.

The Government limits the total number of stock to be grazed on each National Forest. The Supervisor allots the range among the various applicants, giving a preference to the small near-by owner and the men who have always used the range. To drive stock across a Forest it is necessary to get a permit from the nearest Ranger or the Supervisor, except along a public road. A permit is also necessary to drive stock across Forest to reach private lands within it, if the stock grazes along the way.

Get a permit from the Supervisor before constructing drift and pasture fences and corrals.

Remember that most of the range in the west has been overgrazed and that in National Forests it is being brought back into good condition. This means that the Supervisor is often unable to grant applicants the privilege of grazing the full number of stock applied for. It should be born in mind, too, that all the applicants are cut down proportionately, the big man the most. If an applicant has a grievance which he cannot settle with the Supervisor, he may appeal to the Forester (Chief of the Forest Service). On new National Forests existing conditions will not be changed suddenly. Owners of stock will be given ample notice if it is necessary to make a cut, so that they can adjust themselves to the new conditions without financial loss.

Do not graze any stock on a National Forest without a permit (except as noted above) or drive stock across National Forest lands without a permit. The law forbids it. And to guard the best interests of all the people the law will be vigorously enforced.

It was not long, however, before most ranchers realized good conservation on the rangelands worked for their benefit. Many who had initially opposed the Forest Service efforts became ardent supporters.

Thus Gifford Pinchot and his U.S. Forest Service announced that grazing, while permitted in the new National Forests, would be regulated. Depending on the condition of the range, the numbers of stock permitted could be cut back. Users had to get a permit and pay a fee. They had to work closely with the Ranger and Forest Supervisor on range use. Albert E. Cole provides an insight into how well the early Forest Service program was working. Cole left the Forest Service in 1910 and came back in 1917. He wrote:

"I noticed a decided change in the attitude of the general public toward the Forest Service when I returned in 1916. On the Dry Cottonwood District (Deerlodge National Forest) I again came in contact with the Austrians and Italians of the Browns Gulch country, and these people were not nearly so hostile towards the Forest Service as they had been in 1906 when I worked there settling trespass cases. Also I found greater friendliness from the stockmen and other users, and a decided trend of attitude toward cooperation with the regulations that had not been apparent during the earlier days. Stockmen and other users in general had begun to realize that the Forest Service was working for the good of the people as a whole and that the small operator, whether rancher, timber man or sawmill owner, was treated just as fair and with as much consideration as the larger holders. This had a tendency to give the Forest Service a reputation for fair dealing and to make for fuller cooperation and friendliness all around. Also, the fact that the Forest Service had cracked down on some of the large cattle and sheep outfits and had succeeded in enforcing the laws and regulations in numerous trespass cases had engendered a wholesome respect for the Service, and the fact that political pull had no effect on the action of the Supervisors and Rangers in enforcing the law taught the larger outfits that they had to obey these men, and encouraged the smaller holders to a better feeling and a better spirit of cooperation. Also, both large and small operators began to realize that there was real merit in the method of handling stock and timber and that good results were becoming apparent, and a great many of the large as well as the small outfits became actually enthusiastic in following the instructions and orders of the Forest officers."

In 1920 Cole was assigned to the Arcola District, Deerlodge National Forest. Here he was given a range management assignment:

"My first job was to encourage the formation of a livestock association, which was accomplished almost immediately after I assumed charge of the District. The cattlemen were anxious to form such an association, but had been under the impression that the Forest Service would not cooperate."

Working cattle in 1932 at the Van Cleve ranch, Big Timber Canyon on the old Absaroka National Forest.

Sheep grazing in the Beaverhead National Forest in 1943.

Purebred Hereford cattle owned by A.B. Cook on summer range in the Lewis and Clark National Forest in 1927.

Cole was convinced cooperation and coordination between stockgrowers and Forest Service personnel worked well at the local level.

Fred Morrell, Northern Regional Forester from 1920 to 1929, agreed with Cole when he wrote: "There were no unusual problems in the administration of grazing during the time I was Regional Forester." According to Morrell, "____ most of our ranges were in relatively good shape and range administration had been pretty well accepted."

While care and improvement of the range was mutually accepted by those holding grazing permits and Forest Service administrators, fees for grazing livestock have always been fighting words. In the beginning, forage resources were not sold like timber. Timber was appraised according to current market conditions and sold by competitive bidding. Forage, however, was sold based upon a "reasonable fee" to be locally determined according to national guidelines. Forest Service fees were always lower than market value. Gifford Pinchot justified this because he believed small stockgrowers living in or near the National Forests could not compete if they had to pay going rates.

The stockgrowers felt costs should cover only administrative costs. Over the

Purebred steers in pasture at Sappington, Montana.

years the stock growers have managed to exert enough political influence to keep the grazing fees a good bargain—below average rates paid for similar private ranges.

An important part of the public rangelands in the Northern Region are the National Grasslands in eastern Montana, North Dakota, and the northwest corner of South Dakota. Originally homesteaded between 1870 and the 1920s, the acreage limits of the Homestead laws did not recognize the natural limitations of semi-arid lands. Few people during this time realized the critical function of the grass cover on these lands. As a result, millions of acres of originally stable grasslands were ploughed for crops. The crops from the newly turned prairie soils were wonderful—but only during wet years. In the mid-1920s a long drought, with hordes of grasshoppers and high winds blowing away the unprotected soil, led to countless failed farms and mortgage foreclosures. "Home, Home On The Range" for many had turned into "Paradise Lost."

The problem was recognized in 1925 by the Phillips County, Montana, Extension Agent, Henry Lantz. Supported by the Malta Commercial Club, for years he advocated "retiring" the croplands that should retain a grass cover. In 1935, the Congress approved and funded Lantz's plan. Called the "Malta Plan," it established the Resettlement Administration. The purpose of this agency was to buy back the homesteaded and ruined prairie lands. Before long there were active Resettlement programs in all the Great Plains states, Idaho and Oregon. The lands purchased were called Land Utilization project lands, or L.U. lands. In 1936 the Resettlement Administration was transferred to the U.S. Department of Agriculture. In 1937 administration of the L.U. lands was given to the Soil Conservation Service.

During this time, grazing associations were formed that integrated the federally owned lands with livestock operations on adjacent private lands. With technical assistance from the Soil Conservation Service and full cooperation from the grazing associations, the devastated lands were returned to the production of grass. Work done included reseeding, water developments, fencing and control of use by livestock.

In 1954 the L.U. lands in eastern Montana, North Dakota and northwest South Dakota, along with other L. U. lands were transferred to the U.S. Forest Service. In 1958 the L.U. lands in Montana were transferred to the Bureau of Land Management and incorporated into their Grazing Districts. The lands in North Dakota and northwest South Dakota became part of the Custer National Forest. Here the Grazing Associations continue to actively interact with the Forest Service. The National Grasslands in the Northern Region are the Grand

River/Cedar River, located in South Dakota with a few parcels in North Dakota, the Little Missouri in western North Dakota and the Cheyenne in southeastern North Dakota.

There were confrontations. J.N. Templer, retired supervisor of the Deerlodge National Forest tells an interesting tale about Glen Smith when he was in charge of the Custer National Forest when the headquarters were in Ashland, Montana. The story involves trouble Smith had with the Cheyennes. It seems the Cheyennes had a habit of poaching and molesting permitted livestock whenever they rounded up their cattle that had strayed off the reservation. According to Templer:

> "One fall Glen beat them (the Cheyennes) to it, and on the general roundup he saw to it that all the Cheyenne cattle were gathered and thrown over on the reservation. One day after the roundup Glen was in his office when he noticed a string of 12 to 15 wagons with the attendant caviya passing through Ashland. Smelling a mouse, Glen halted them and asked Redbird, the Indian in charge, where they were headed for. He was told the outfit was going up on the Forest Reserve to gather the Cheyenne cattle. Glen replied that he had gathered all their cattle and that they had been thrown onto the reservation. However the Indians insisted that not all of their cattle had been gathered and that they were going to do that little job with precision and dispatch. Thereupon Glen acquiesced, but told them that he was going to search their wagons for guns, which he did, going from wagon to wagon without result until he came to the last wagon. There he found a small arsenal and, with his color rapidly rising and his suspicions confirmed, Glen forcibly informed them they could not go on the Forest unless they left their guns at his office.
>
> This brought things to a head and the Indians, refusing to leave their guns with Smith or change their intentions, started down the road. Glen straightway mounted his horse, and with a 30-30 started to head off the Indians (he having been delayed a bit in going after said horse). He circled and met the outfit about two miles out of Ashland, and upon stopping them he again told them they could not go on the Forest with their firearms. Apparently this did not deter them in their intent to continue on their way, and Glen, seeing that force would probably be necessary, then and there declared himself. In language, whose forcefulness left little to be desired, Glen told them that the only way they could go on would be over his dead body. This put a different light on the matter. After powwowing a while with much gesturing towards Glen's body, the bunch gave in and faced about. Local people say that years later the Indians let the cat out of the bag by saying that the way Glen handled his 30-30 did not deter them, but the thought of having to surmount Glen's body just to kill a few deer was a little too much."

Managing the range required long hours in the saddle, helping move and drive livestock, facing the bright sun and the driving endless wind. It meant getting soaking wet, forever eating meals of sidepork and beans washed down with scalding coffee. On a good day there were canned peaches for dessert. You learned to roll your own cigarettes and take care of your horse before you took care of yourself—no matter how cold, wet or tired you were. A pouch formed in your lower lip where you put your snoose. It was around countless evening campfires that most of the important decisions and agreements were made concerning how the range should be managed.

The life of Rangers in cow-country required a cool head to make good decisions and a strong hand to make them stick.

It was harmonicas, guitars and the tinkling sounds from a bell mare that provided background music. These were the elements for creation of those magic moments when the night sky, earth, cowboy and ranger blended into a single entity. It was the time when range problems were easily resolved because they were, after all, by comparison, insignificant.

Sheep Driveway

By Lee Hames

One of the largest sheep grazing areas in western Montana was the mountains between the Bitterroot River Bridge south of Missoula and the Idaho line at Lookout Pass. Starting in the early 1900s, sheep were trailed into this area by the thousands. A trail, never really planned or surveyed, evolved as sheepherders drove sheep in and out of the mountain ranges under public domain. This trail became known as the Sheep Driveway. In many places the driveway followed old trails formerly used by Indians and trappers, and in other places, new trails developed as grazing pushed farther into the mountains.

The Forest Service, in addition to timber management and fire prevention and control, regulated livestock grazing on the public domain by issuing grazing permits. They always placed cattle and sheep on different ranges. Lands with milder terrain at lower elevations were used for cattle grazing, while higher lands with steeper, alpine-like terrain were delegated to the grazing of sheep. Cattle ranges were usually adjacent to or as near as possible to the home ranch. The Forest Service controlled the number of sheep allowed to graze and set dates for interring and leaving particular ranges. They allotted every band of sheep (1,000 ewes plus their lambs, approximately 2,250 head) a designated area in which to graze for the summer season. When ownership of a ranch changed hands, grazing permits were usually transferred to the new owner.

One camp-tender and one sheepherder controlled and protected a band of sheep. The camp-tender cooked, put salt out where the sheep would bed down for the night, and moved the camp when necessary. Since the Forest Service allowed the herd only three nights in the same place, the camp-tender had to locate a new bedding ground and move camp often. Because of this constant moving, the camp-tender kept pack animals with him at all times to carry salt boxes and sacks of salt, tents, bedrolls, food, and all the equipment necessary to maintain the camp. Each new campsite had to have level ground for the sheep to bed on, a watering hole big enough for both sheep and pack animals, plus water suitable for drinking and cooking. Thus, the camp-tender had to be a very capable person and spend many hours at his work.

The herder never worked around the camp. His sole responsibility was to watch the sheep at all times. Every morning at day light, the sheep would leave the bedding ground and start grazing. The herder and his dogs always left with them. The sheep usually grazed until about noon when they would have a drink of water before seeking the shade for an afternoon nap. The herder and his dogs napped with them. About four in the afternoon, the sheep would leave the shade to resume grazing until dark. The ringing of the little bells on their necks would wake the herder and his dogs. The herder, then, coaxed the sheep to graze in the direction of the camp where they would bed down for the night. The sheep stayed in the mountains from June until October. During this time, neither the herder nor the camp-tender had a day off.

Since the sheep ranges were usually far back in the mountains, pack strings had to bring them all their supplies. About once a week, a pack string from the outside arrived at the sheep camp with fresh meat and vegetables, groceries, and salt for the sheep. The sheepherder and camp-tender welcomed mail, magazines, and newspapers with great enthusiasm.

All told, however, if a person didn't mind solitude and loved to be out in the mountains, surrounded by the most beautiful scenery imaginable, sheep herding was an idyllic existence. Life in a sheep camp was not all that bad. It was sort of a lazy man's job. It didn't require heavy lifting. You just sat on a hillside and watched the sheep graze. There are many city dwellers who would envy the peaceful life of a sheepherder. Many of the mountain streams were filled with trout. Every day, you could catch sight of deer, elk, bear, and the occasional moose. Sometimes the bears might try to harm the sheep, but most often the presence of the herder and his dogs kept them away.

During the Civilian Conservation Corps (CCC) days of the 1930s, there was a CCC camp at the head of Graves Creek near Lolo Hot Springs. One of the camp's jobs was to clear brush from a pathway, about 50 feet wide, to define the Sheep Driveway running along the tops of the ridges. The main route of the driveway started on the east end of the sheep ranges in the Bitterroot mountains, directly across the Bitterroot River from Fort Missoula, near the Bitterroot River Bridge. Going over Blue Mountain, the driveway stayed on top of the ridges running parallel with Lolo Creek to Wagon Mountain and then on to the Idaho line. The driveway passed between the headwaters of Petty Creek and the headwaters of Granite Creek and the South Fork of Fish Creek. Throughout this section of the Sheep Driveway, trails branched off to the north that went to Diamond Point, Tank Creek, Albert Creek, and the South Fork of Fish Creek. Bands of sheep followed these trails to their summer ranges.

North of where present-day Highway 12 crosses the Montana-Idaho state line, the driveway forked. The left fort went to Packers Meadow, Elk Meadows, and the Brushy Fork. The right fork followed the old Indian trails west along the state line. This fork of the driveway passed near the South Fork of Fish Creek, Thompson Creek, Surveyor Creek, Cedar Log Lake, and Fish Lake. The right fork of the Sheep Driveway then curved around the headwaters of the North Fork of Fish Creek and Straight Creek to Goose Lake and the HooDoo Meadow and Pass area. From HooDoo Pass the driveway went west along the state line to Lookout Pass, Kelley Creek, and the Little Joe River. All together, the driveway was about 100 miles long, not counting the trails that branched off of it. The Sheep Driveway became one of the few trails leading into the Bitterroot mountains.

To the north of the Sheep Driveway, many long, open ridges, covered with grass, extended from the driveway to the Clark Fork River. Herders drove many bands of sheep over the driveway to get to these grassy ridges.

One of the first men to trail sheep over the driveway before the CCC cleared the way was D.R. Maclay of Lolo, Montana. Leaving his range east of the Bit-

Packer arriving with supplies for the summer sheep range.
COURTESY JACK UNDEM

terroot River, D.R. drove his sheep across the Bitterroot River Bridge and over Blue Mountain to the head of Woodman Gulch. He took his herd down Woodman Gulch, across the Lolo Creek Road, and up the South Fork of Lolo Creek to Elk Meadows. The herd summered in Elk Meadows, Packers Meadow, East Fork of Lolo Creek, and the Brushy Fork. Pack strings brought in supplies on a weekly basis. Over the years, as D.R. increased the size of his herd, he pushed further into the mountains until he and his herders had trailed them as far as Kelley Creek and the Little Joe River.

When D.R. went out of the sheep business, he leased his sheep range east of the Bitterroot River, between Davis Creek and Eight Mile Creek, to Louis Undem who continued to use it for a sheep operation. Undem usually kept three bands on the range. He wintered, lambed and sheared east of Florence at the headwaters of Eight Mile Creek. During the summer, Undem and his herders drove the sheep over the driveway to various summer ranges in the mountains. The last year Undem used the driveway was 1954. The summer of that year, he had one band in the area around Petty Mountain, Wild Horse Mountain, and Diamond Point. Another band grazed the area around Fish Lake, Cedar Log Lake, Surveyor Creek, and Thompson Creek. After that season, Undem went out of the sheep business and bought into the cattle business. I believe Undem was the last person to drive his sheep over the driveway from its east end.

Sheep shearing at Antvines ranch, 1940. LEE HAMES PHOTO

Stanley Antrim had land on the Eight Mile Flats, adjoining Undem's land, where he also operated a sheep ranch. Antrim always ran at least two bands, and some years he had three. His operation was similar to Undem's but, unlike Undem, he sent only one band over the driveway each year. Antrim's herders would drive the band to the West Fork of the Bitterroot and summer the sheep near Bare Cone and Castle Rock lookouts. The last few years Antrim was in the sheep business, he shipped his herd to Browning where he had obtained summer grazing rights from the Blackfeet Indians.

W.A. Simons, who owned the Wilma Building in Missoula, also owned a ranch directly across the Bitterroot River from Fort Missoula. At this ranch he kept two bands of sheep for wintering, lambing and shearing. He moved his sheep to several thousand acres in the O'Brien Creek area for spring grazing. As soon as he could, each year he moved the sheep into the mountains for summer grazing. Billy Seifert, his ranch manager and foreman, trailed two bands over the driveway in 1941. Some time during WWII, Simons went out of the sheep business and sold his property near the river and O'Brien Creek.

Sheep shearing at Eight Mile Flats in the Bitterroot Valley, 1940. LEE HAMES PHOTO

During the 1920s, Dave Graham had a sheep operation on the hills south of Missoula that are now covered with houses. He had a small operation, but he did use the driveway to get to his summer range.

Dr. McCullough owned the Miller Creek hills, known now as the KO Rodeo Ranch, and wintered, lambed and sheared at his Miller Creek ranch headquarters. McCullough was only in the sheep business for a few years during the 1920s, but he too, pushed sheep over the driveway for summer grazing.

Nick Lasko lived north of Hot Springs near the Nirada Post Office. He had a tremendously large wooden sheep shed close to the highway where he wintered, lambed and sheared several bands of sheep yearly. In the summer, Nick trailed his sheep through Hot Springs to Perma, then to Paradise and through the cutoff to St. Regis. Near St. Regis, he entered the mountains and got on the driveway. From this point, he drove on to Kelley Creek and the Little Joe River.

About four miles east of the Lasko place, Dan Polasson had a sheep operation. He followed almost the same winter-summer pattern as Lasko. These two men were Rumanians who had emigrated to the United States together as boys and had worked with sheep since their arrival. By the late 1950s and early 1960s, both men had reduced their operations to the number of sheep they could graze nearer to home; then, they both stopped using the driveway.

Between Perma and Dixon, north of the Flathead River, a very large sheep outfit known as the Mahoneys owned thousands of acres of land. They sent thousands of sheep over the driveway into various summer ranges. They got out of the sheep business in the early '30s and the Forest Service leased their land for a place to winter horses and mules.

Up the Flathead River from the Mahoneys, Nye and Small had a ranch where they wintered several bands of sheep. During the summer, they trailed through the Paradise cutoff to grass along the Montana-Idaho state line. They too left the sheep business and sold the ranch. It is now being used to raise cattle.

Another group of sheepmen entered the driveway at the west end near Lookout Pass. They were Basque people who originally came to the United States from the Pyrenees Mountains on the border between France and Spain. The Basque people were known the world over for their ability to raise and handle sheep. The Basques wintered, lambed and sheared around Moses Lake, Quincy and Ritzville, Washington, and spent their summers on the Montana-Idaho border. The warmer climate in Washington allowed them to lamb around the first of the year. So, by the time his herd was shipped to Taft and Superior in Montana, the lambs were old enough to trail with their mothers from the railroad to the summer range.

Bays of wool LEE HAMES PHOTO

Starting in the early 1900s, five Basque sheepmen—Martinez, Martinella, Gariboldi, Seveille and Fjoure—shipped their herds into Montana for summer grazing along the driveway. Not all of them came every year, however, so it is hard to give an exact count of the number of sheep they ranged. But, there were always at least half-a-dozen or more Basque bands on the driveway.

The last of the Basques to come out of Washington was John Fjoure from Quincy. During the 1950s and '60s, Fjoure shipped as many as four bands into Taft, St. Regis and Superior, where he had facilities for onloading. My partner and I would take our livestock trucks to these onloading point in the spring, and haul Fjoure's sheep into the mountains as far as the large semis could go. Fjoure would set up portable chutes for offloading and then trail his sheep onto the range. In the Fall, we would return to the loading chutes to haul his lambs to the railroad to the east for marketing. Fjoure would tail his ewes from the mountains to the railroad and then ship them back to Quincy for wintering. We hauled Fjoure's sheep for quite a number of years. The last year John Fjoure came to the Superior-St. Regis area was 1966. He only brought two bands.

Once, when John and I were settling our account, he told me that he had come to this range as a young herder and had been there every summer for the last 58 years. He and Tony Seville had started in the sheep business as herders on this range. They worked hard and saved until they built ranches of their own.

By 1966, John was getting too old to get around in the mountains. He decided that this would be his last year on the driveway, and that Fall he shipped his lambs and one band of ewes to market. He took the other band of ewes to Quincy where he had enough range to run them.

At one time, Montana was a very large sheep producing state. But, for various reasons (the shortage of good, reliable sheepherders for one), most of the big operations have disappeared from the scene. With the exception of a few larger operations and some farm flocks, the sheep business is another industry that has slipped away from Montana.

Sheep grazing near Blue Mountain, Lolo National Forest. COURTESY JACK UNDEM

CHAPTER SEVEN:
Recreation

A piece of rainproof canvas, some poles, a parklike stand of trees in a National Forest, an automobile and a willing family are the ingredients needed for camping. Perhaps this simple tent was inspired by the tepee lodge of the greatest campers in North America—the Indians of the Great Plains.

Recreation

Outdoor recreation in the Northern Region was not a very important factor prior to the end of World War II to the Forest Service. In fact, during this period, most people would smile or laugh at a suggestion that the most important use for millions of acres of National Forest lands would someday be outdoor recreation. In the early years, while most local people made extensive use of National Forests, they did not think of this use as recreation—to them it was subsistence.

Hunting, fishing, trapping, picking huckleberries, cutting firewood—these were tasks that needed doing as part of making a living in the West. They were not activities you did for the fun of it. Doing them may have been enjoyable but, how ridiculous to even imagine someone *wanting* to hike 50 or 100 miles for *pleasure.* It would have been impossible to convince very many of these people that universities would one day offer degrees to students majoring in "Outdoor Recreation" or "Leisure Sciences." This was something fiction writers would dream up.

The 1907 publication "The Use Of The National Forests" barely mentions outdoor recreation. Under the section "TO USE WELL ALL THE LANDS," it states:

> Playgrounds—Quite incidentally, also, the National Forests serve a good purpose as playground for the people. They are used more or less every year by campers, hunters, fishermen, and thousands of pleasure seekers from the near-by towns. They are great recreational grounds for a large part of the people in the West, and their value in this respect is worth considering.

There was, however, one very important group, outfitters and guides, that existed before the turn of the century—long before the Forest Reserves were created. They are still a very important part of the outdoor recreation in the

Maps to answer tourists questions about, "Where am I?" "Where can I go?" and "What can I do?" have always been important. Here is an early Forest Service map sign showing "The Recreation Realm of the National Forests in Eastern Washington, North Idaho and Montana."

Early efforts to attract tourists often included excellent accommodations within the National Forests, such as this club house and hotel at Hayden Lake in the Coeur d'Alene National Forest.

Natural geologic features such as hot springs have always attracted visitors and developments. Here are some early-day improvements at the Red River Hot Springs in the Nezperce National Forest.

Northern Region. Perhaps the first outfitters and guides were those serving hunters who came into the area of the Northern Region to kill buffalo and other big game. Good hunting and fishing, today as in the past, continues to attract new customers and to bring back repeat business.

By the early 1930s dude ranches were beginning to gain in popularity. In a 1931 brochure, the "Lone Wolf Ranch" advertised, "Enjoy real western ranch life, cowboys, broncho riding. Rest and rebuild mind and body in Montana's magnificent mountain climate. Trout fishing and pack trips, big game hunting in the Fall." Every year more people were relaxing from the pressures of living in populated regions by visiting the undeveloped areas of the West. They were finding out that after such a visit they could return to familiar pressures renewed and refreshed in body and spirit.

During most of the period covered by this history, railroads were the principal transportation to and within the Northern Region. After automobiles became practical, and as roads were constructed to open up areas that had been inaccessible, recreational use in the National Forests increased dramatically.

Other important influences on outdoor recreation in the Northern Region were Yellowstone and Glacier National Parks. These are two of the greatest

Something for supper! Here is the result of a 1937 visit to Hoodoo Lake in the Selway Bitterroot Primitive Area, Lolo National Forest.

There are times and places reflection includes mountains and trees in the water as well as an inward look into your soul—as this fisherman has discovered in 1924 at Upper Twin Lake, Deerlodge National Forest.

tourist attractions in the world, and are almost encircled by National Forests. Again as transportation improved, these two National Parks attracted multitudes of people. Many came from the East. Many also "discovered" the National Forests of the Northern Region.

By 1935, the importance of outdoor recreation was reflected in a brochure for the Helena National National Forest. It states:

> RECREATION
>
> Of all the uses of the national forests and their many resources, recreation is the most popular. Recreational use, which includes camping, hunting and fishing, hiking, saddle and pack trips, swimming, motoring, and many other outdoor activities, increased from 3,000,000 people in 1917 to more than 34,000,000 on the National Forests in 1933. Many people lease, for a nominal sum, sites for summer homes near beautiful lakes or beside rushing streams. [Author's note: New leases are no longer available.] Others stop at public camp grounds, most of which are carefully planned and developed. Primitive areas, maintained within the National Forests, without roads and with no human habitations except for an occasional forest ranger station or lookout tower, offer unsurpassed recreational opportunities for those who really wish to "rough it."

Perhaps the single most important chain of events, shaping the future of outdoor recreation in the Northern Region, culminated during this period. This was development and implementation of the concept of "wilderness." Thoreau, one of the first to recognize the value of wilderness, wrote: "In wilderness is the preservation of the world." John Muir, founder of the Sierra Club, added: "In God's wilderness lies the hope of the world—the great fresh, unblighted, unredeemed wilderness. The galling harness of civilization drops off, and the wounds heal ere we are aware." These were powerful words. They offered hope to millions who felt they were burdened by the "galling harness of civilization."

Riding horses for pleasure has been important in the National Forests since their creation, as this group is doing in 1937 in the Seeley Lake District, Lolo National Forest.

Next, two professional foresters, Aldo Leopold and Robert Marshall, successfully waved the wilderness banner. Leopold, who once was a Forest Service employee, said, "Many of the attributes most distinctive of Americans are due to the impress of wilderness and the life that accompanied it." Robert Marshall, who spent much of his brief life in the Northern Region, organized the Wilderness Society in 1932 while he was an employee of the Forest Service.

The accomplishments of Bob Marshall, who died suddenly when he was only 38, are legendary. He was born in 1901 to a wealthy Manhattan family. In this respect, he had a common bond with Teddy Roosevelt and Gifford Pinchot, who were also from wealthy eastern families. Marshall graduated from the New York State College of Forestry and came to the Forest Service in Missoula in 1925. Here, before long, he became known for his 50 mile hikes—on weekends, just for the fun of it! By the time he was 36, Marshall had made over 200 hikes that covered 30 miles a day, over 50 that covered 40 miles and several of 70 miles. He wrote:

> "Toting a 50-pound pack over an abominable trail, snowshoeing across a blizzard-swept plateau or scaling some jagged pinnacle which juts far above the timber all develop a body distinguished by a soundness, stamina and elan unknown amid normal surroundings."

Perhaps part of Marshall's acute sense of wilderness and wilderness values were the result of his urban upbringing which contrasted with the background of the vast majority of Forest Service employees. Most were from rural areas and their families were not wealthy. A year before his death, Marshall wrote:

> "We're all young enough that we'll probably meet many defeats in the next 50 years. It's even conceivable that when we die we still will not have won the fight. But win or lose, it will be grand fun fighting and knowing whatever we do in the right direction will help eventual victory."

Marshall did not subscribe to Gifford Pinchot's philosophy of "the greatest good to the greatest number in the long run," applying to everything. According to Marshall:

Bob Marshall, a renowned backcountry hiker and the father of the National Wilderness System, left thousands of footprints in remote areas of the National Forests in the Northern Region.

> "If it did, we would be forced to change our metropolitan art galleries into metropolitan bowling alleys. The library of Congress would become a national hot dog stand, and the Supreme Court building would be converted into a gigantic garage where it would house a thousand people's autos instead of Nine Gentlemen of the Law."

Marshall was personally responsible for adding 5.4 million acres to the Nation's Wilderness System. In 1941, two years after Marshall's death, the Bob Marshall Wilderness Area, an area of almost a million acres, was set aside in western Montana. It was the ultimate legacy of Marshall, that through the concept of wilderness, he should influence the pattern of primitive outdoor recreation that has made the National Forests of the Northern Region world famous. It is only through the establishment of an extensive and enduring Wilderness System that this form of outdoor recreation can be perpetuated. It is only through Wilderness that future generations can actually share the experiences that thrilled Lewis and Clark and other early explorers of the Northern Region.

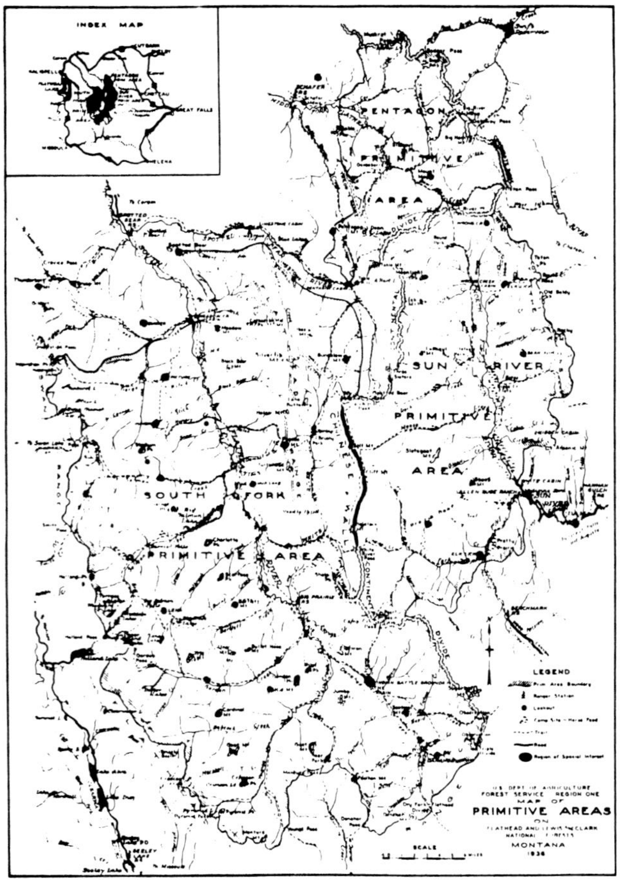

Bob Marshall at the Priest River Experimental Station, 1928. UM MANSFIELD LIBRARY

Everyone, young and old, loves a picnic, and the National Forests of the Northern Region have always offered great places to have them. This 1936 picnic was held at Echo Lake in the Deerlodge National Forest.

Camping in the Tobacco Root Division, Gallatin National Forest in 1936. The design of the tents looks like many in use today. Note the picnic table provided by the Forest Service.

Mr. and Mrs. Fred Hornaday in camp in 1946 at Camp Creek, Flathead National Forest. Note the use of canvas and poles to form a tepee is still popular.

"Packing in" means getting to beautiful remote places seldom visited in the course of a year. This gem is Mirror Lake in 1936 in the Spanish Peaks Primitive Area, Gallatin National Forest.

Riding and packing into remote corners of the Northern Rocky Mountains.

A climbing party in 1946 crossing a snowbank at the Chinese Wall in the Bob Marshall Wilderness Area, Lewis and Clark National Forest.

One of the most important groups on a trail ride is the cook outfit. Here they are traveling toward a camp near the Chinese Wall, Bob Marshall Wilderness Area, Lewis and Clark National Forest in 1946.

The great waterways that served as transportation routes in the area of the Northern Region have always been popular with recreation visitors. Here in 1939 is an excursion boat at Hilger Landing on the Missouri River, Helena National Forest. This craft took visitors to the "Gates of the Mountains," a place noted by Lewis and Clark on their Voyage of Discovery in 1805. The name is applied to an area where the mountains on either side of the Missouri seem to form an impenetrable wall that opens like a gate as it is approached from the river.

Trail riders in 1946 in the meadow below the Danaher Ranger Station, Flathead National Forest.

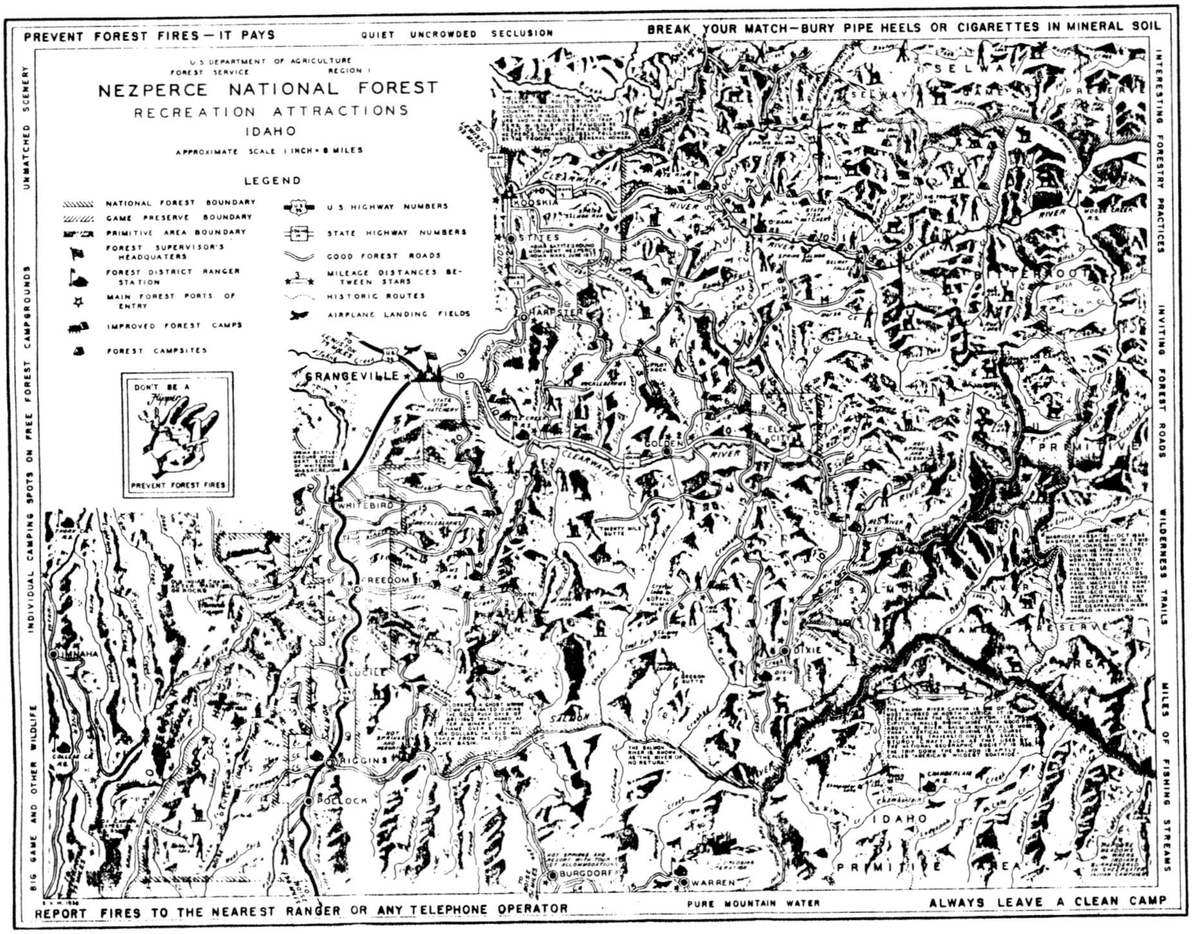

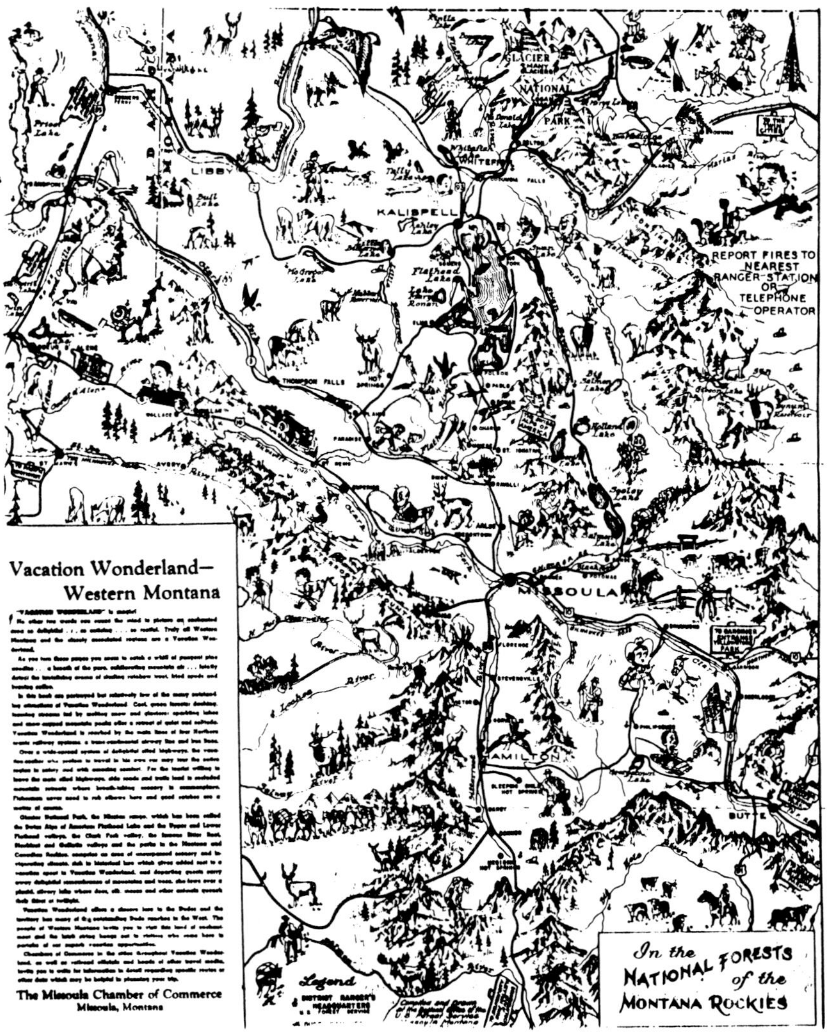

Vacation Wonderland— Western Montana

The Missoula Chamber of Commerce
Missoula, Montana

Winter Sports

The sport of downhill skiing in the United States goes back to the mid-1930s. Most of the early ski areas were established on Forest Service land. There were a number of small areas in Region One developed from the 1930s to the beginning of World War II. Some areas had rope tows of various lengths, some had only a few cut runs. A few had small ski lodges or warming huts built by the Civilian Conservation Corps.

A 1941 booklet lists the following ski areas on forest service land: Chewelah Hill and Butler Mountain, Kaniksu National Forest; Lookout Pass, Coeur d'Alene National Forest; North and South, St. Joe National Forest; Hellroaring and Belton, Flathead National Forest; Pattee Canyon, Lolo National Forest; Gibbons Pass, Bitterroot National Forest; Silver Lake, Blue Grouse Hill and Baggs Creek, Deerlodge National

WINTER SPORTS
IN THE
NORTHERN REGION
NATIONAL FORESTS

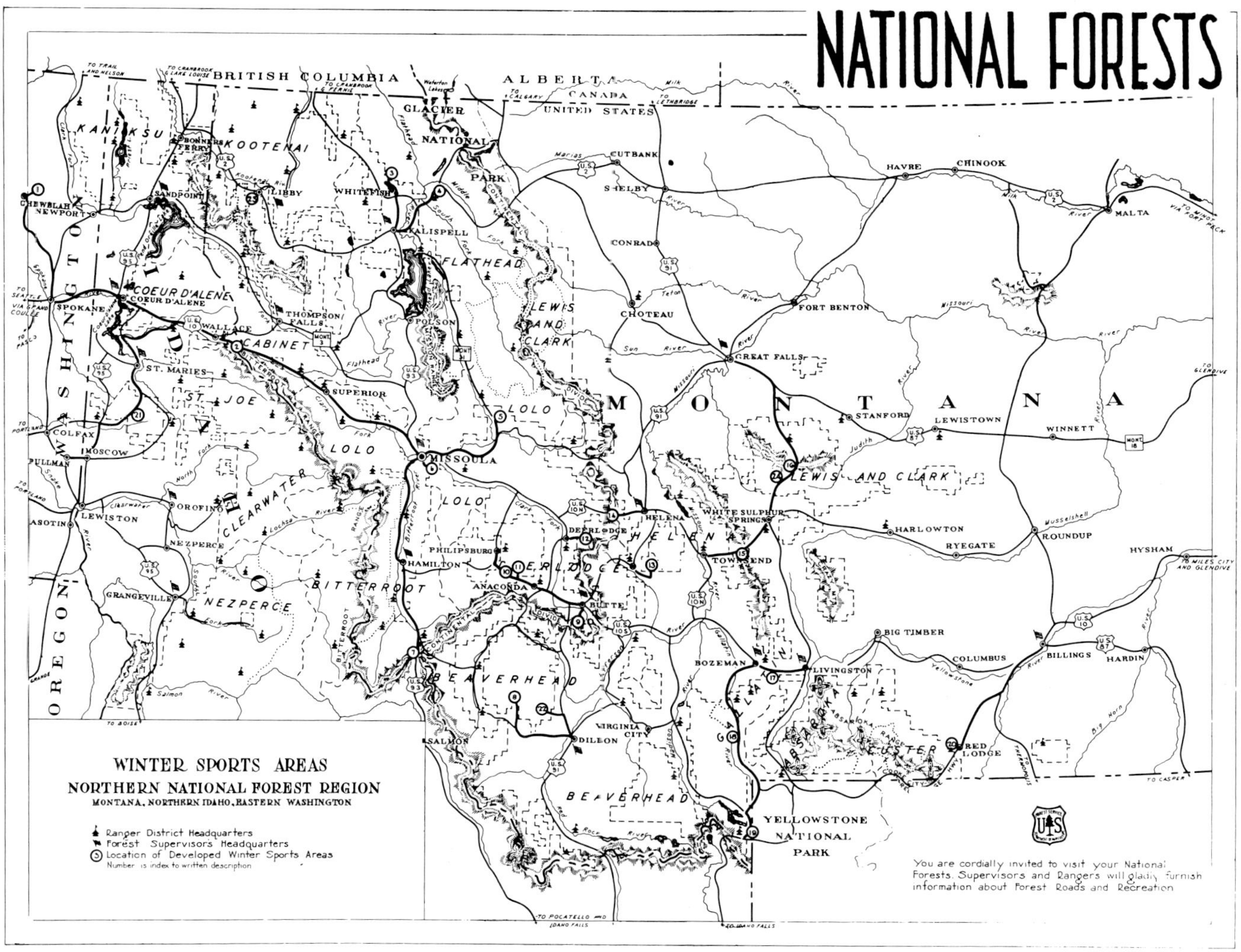

Forest; MacDonald Pass, Helena National Forest; Kings Hill and Forest Green, Lewis & Clark National Forest; Karst's and West Yellowstone, Gallatin National Forest; Willow Creek, Custer National Forest; Libby, Kootenai National Forest.

The present ski lodge at Lookout Pass and part of the lodge at the North-South Ski Bowl, both in Idaho, were built by the CCC.

The war would curtail any further development of the areas and it would take some returning veterans of the famous 10th Mountain Division at the end of the war to spark an interest in massive development of areas in Region One and throughout the country.

Ski lodge at Lookout Pass on the Montana-Idaho border. Constructed by the CCC in 1941. Part of the ski lodge at the North-South Ski Bowl near St. Maries, Idaho, was also constructed by the CCC.

Idaho skiers, early 1900s. COURTESY IDAHO HISTORICAL SOCIETY

Skier on the Continental Divide, Jan. 9, 1935.

CHAPTER EIGHT:

Transportation

By the late 1930s, roads were being improved to take advantage of better, faster motor vehicles. This highway near Neihart, Montana, in the Lewis and Clark National Forest is a good example of a modern road in 1937.

An early gas shovel and truck building a new road in the Gallatin National Forest.

Road on the east side of Riley Pass, North Caves Hills Division, Custer National Forest in South Dakota.

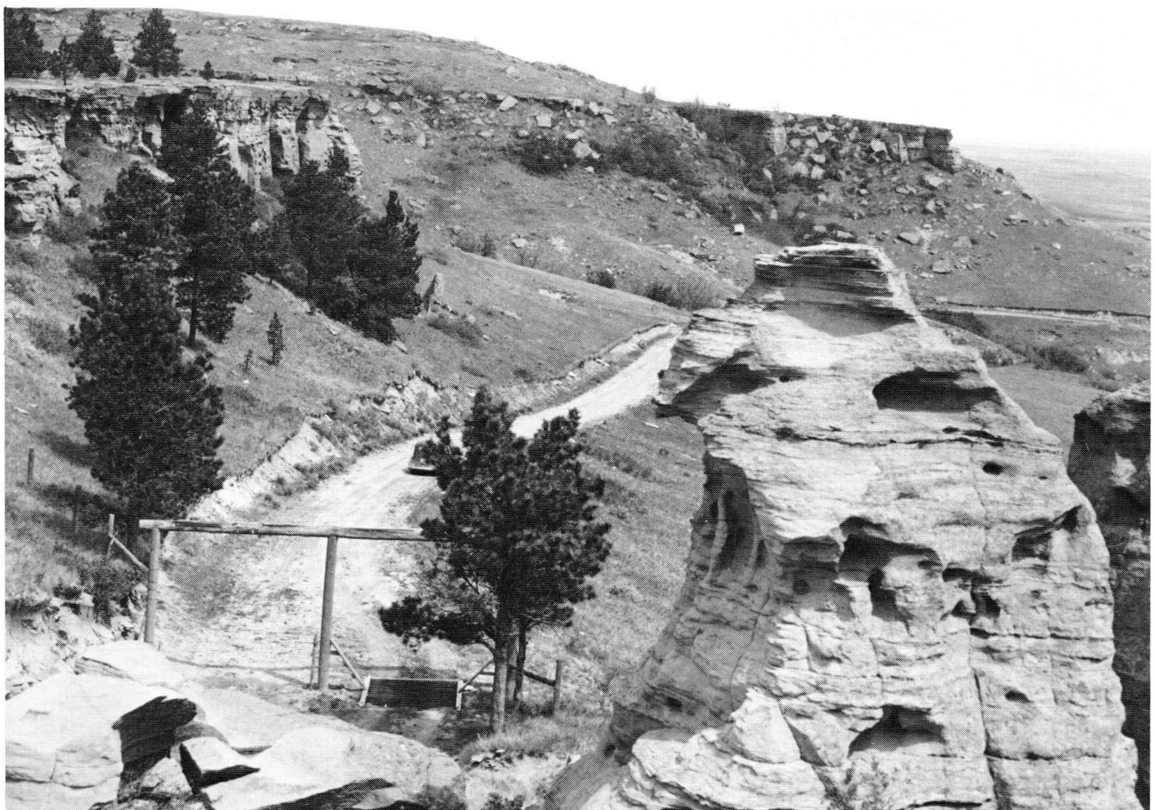

Highway near Neihart, Montana, in the Lewis and Clark National Forest, 1937.

The evolution of the automobile and the evolution of roads to accommodate them is demonstrated by this road in the Madison Canyon, Gallatin National Forest.

Thousands of miles of roads were constructed under the Civilian Conservation Corp. This is the Salmon River at the "Crevice" on a CCC constructed road in the Nezperce National Forest.

King's Hill Pass on September 9, 1916.

The road between Troy and Libby, Montana in 1915, Kootenai National Forest.

Snow on Lookout Pass at the boundary between Montana and Idaho.

Lolo Pass

Walls of snow on the Lolo Pass, Powell Ranger Station.

May 30, 1933, on the Lolo Pass. The horses are Diana, Joe, Reilly and Tex.

More snow on Lolo Pass.

A narrow road over shale rock east of Fourth of July Canyon, Idaho Panhandle National Forests, 1916.

The road over 4th of July Pass before the construction of Interstate Highway 94.

View of the log construction on the Gallatin Highway at the entrance to the Gallatin National Forest, 1937.

This is Highway 10, now replaced by Interstate Highway 94, at the "Camel's Hump" in the old Cabinet National Forest in 1938.

A Forest Ranger on a developed trail in the St. Joe National Forest in 1920. In these days trails were the backbone of the transportation system.

Early automobiles were designed to use wagon roads such as this in the Mission Mountains of the Flathead National Forest. With their flat springs and high narrow wheels, they resembled horse-drawn buggies with a gasoline engine. Before long, however, they would replace horses and roads would be designed for faster improved motor vehicles.

Silent witnesses to a wildfire serve as sentries in 1938 along the Lolo Divide road near Castle Butte in the Clearwater National Forest.

Pack Strings

If you could fit it on a pack saddle, there were mules to move it. Here lumber is being loaded at the Honeysuckle Ranger Station, Coeur d'Alene National Forest in 1932.

Ranger Chester Jackson near the head of Rainbow Creek in the old Absaroka National Forest in July, 1932. The pack horse is carrying the equipment of the famous Forest Service photographer, K.D. Swan.

Burros were also tried in the early days of the Northern Region. These hardy animals were loaded with supplies to deliver to a trail maintenance crew in 1931, Flathead National Forest.

A pack string loaded with "shake blocks" in 1929. These are blocks of cedar that will be split into shingles after they have reached their destination in the Lolo National Forest in 1929.

Loaded with supplies brought in a plane from Missoula, a pack string leaves Bartlett Creek Landing in 1938.

In the 1920s, trails, horses and mules were literally the backbone of the transportation system used to move equipment and supplies.

In 1939, a Ranger from the Bonners Ferry Ranger Station is ready to start on a field trip. He is combining the best of two worlds—a pickup truck and trailer to drive to a trail head, and a horse to use on the trail.

Pack string being loaded in 1932 at the Honeysuckle Ranger Station in the old Coeur d'Alene National Forest.

Moving fire fighting equipment was important and often dangerous work. Pack strings getting tools and supplies to the right place at the right time often was the difference between stopping a fire or its going on a rampage over thousands of acres. This pack string is on its way to a fire in the Lolo National Forest in 1937.

CHAPTER NINE: Ranger Stations

H.C. Tuttle, left, and N.E. Wilkerson, right, are believed to have constructed this first Forest Service Ranger Station and floated the first flag over a Forest Service building at Alta, Mont., in 1899.

The sign reads:
"OLD ALTA RANGER STATION"
This was the first U.S. Forest Service Ranger Station. It was built in 1899 by H.C. Tuttle and Than Wilkerson.

In 1904 this site reverted to private ownership under a mining claim. The Lion's Club of Hamilton, Mont., purchased the site in 1941 and donated it to the Forest Service for preservation as a historic Montana landmark.

The New Falls Ranger Station, Kaniksu National Forest under construction.

Hanon Ranger Station, 1908-09

Exterior view of the Hanon Ranger Station, Lolo National Forest in 1909.

The kitchen in the Hanon Ranger Station, Flathead National Forest in 1909.

A place to sit and eat in the Hanon Ranger Station in 1909. Note the display of pin-ups on the wall along with a rifle, pistol and canvas water bag.

An early picture of the Forest Service Ranger Station at St. Regis, Mont.

The Forest Service Ranger Station at Seeley Lake, Lolo National Forest in 1910.

Another early view of the Ranger Station at Seeley Lake, Mont.

The Bonita Ranger Station Office Building, Lolo National Forest, in 1937.

An early view of the Forest Service Ranger Station in Wallace, Idaho.

The Slate Creek Ranger Station, Nezperce National Forest in 1923.

Adams Camp and Ranger Station, Nezperce National Forest in 1925.

Another view of Adams Camp and Ranger Station in 1925, Nezperce National Forest.

Montour Ranger Station in 1925.

An early view of the McGruder Ranger Station.

The Pete King Ranger Station, Selway National Forest in 1921

The Powell Ranger Station, Clearwater National Forest in 1931.

Shoveling snow off the roof of the Powell Ranger Station, Clearwater National Forest, in 1935.

Northern Regional Forester, Major Kelley, right, on a visit to the Powell Ranger Station in the 1920s.

An early view of the Lolo Hot Springs Ranger Station, Lolo National Forest.

The Forest Service cabin at Medicine Springs Ranger Station, Bitterroot National Forest in 1930.

State Line cabin on the Lolo Pass

Interior of the Ranger's Office at Kelly Creek Ranger Station, Clearwater Nastional Forest in 1932. The boards used here were all whip sawed, including the table, windows and door frame. Part of the equipment mounted with the switch board was used to call this station with an Adams portable telephone.

An early picture of the West Yellowstone Ranger Station, Gallatin National Forest.

The Fort Howe Ranger Station. Note the scoria roofs.

Fenn Ranger Station

Fenn Ranger Station, the administrative center for the Selway Ranger District of the Nez Perce National Forest, was built during the years 1937-1940. Built at a cost that was staggering for the time, $500,000, the station was one of the most elaborate built by the Forest Service, and the attention to site design and building placement had not been seen in Forest Service facilities prior to this. The extensive attention given to the design of the station can be attributed, at least in part, to the use of the Civilian Conservation Corps.

The final design was created by Northern Region architect William Fox, who has other Ranger Stations and buildings to his credit on the National Register of Historic Places. Materials, stone and cedar, were procured locally. A quarry was located nearby, and split cedars for the roof were made at a mill near the station. Although the buildings were resided in the 1950s because of weathering problems resulting from the damp climate, the overall style and the details and design of the station have been retained.

The CCC enrolles who built Fenn were stationed at the O'Hara camp, built in 1935 a short distance from the station.

The location chosen for the station, a sloping bench, facing south, above the Selway River, accentuates the attractiveness of the buildings. The station is set back from the road and is surrounded by maintained lawns and trees and sweeping curved drives paved in 1986. Rail fences surround the adjacent pastures. Originally called Goddard Bar Ranger Station, historic records state the name was changed in 1939 to honor the first Nez Perce National Forest Supervisor, Major Fenn. Only minor changes have occurred to the station or its setting since 1940.

The range station consists of 15 buildings, 8 structures and 2 objects: administrative building, cookhouse, bunkhouse, east warehouse, west warehouse, gas house, west garage, barn, pumphouse, gasoline center, east residence, west residence, garage for residences, a lower west residence, and garage, weather station complex, helicopter landing pad, four satellite dishes, an entrance sign and a stone wall supporting a series of interpretive signs.

CHAPTER TEN:

Region One Headquarters

Loading fire equipment into trucks in 1920, at Missoula, Montana. Note the fire message, "HELP PREVENT FOREST FIRES." This is still the basic theme of most fire prevention messages.

Interior of the Forest Service Northern Region warehouse in 1920 which was located at 240 West Pine St., Missoula, Montana.

A Forest Service office in the Lolo National Forest headquarters in the early days. This is believed to be taken in the Hammond Arcade Building in Missoula, Mont. Note the clean desk in front and the back tables piled with paperwork.

An early Forest Service "work station" in the Lolo National Forest Supervisor's Office. Note the long desk and row of hanging rubber stamps used for making entries in ledgers like the one open in the lower left. Also note the "Smokey Bear" hat hanging on the wall.

An early-day meeting in the Lolo National Forest Supervisor's Office in the Hammond Arcade Building, Missoula, Mont. Maps mounted on the wall, a telephone (which looks recently installed) manuals in a bookcase and a mound of paperwork appear to be standard ingredients for many Forest Service get-togethers.

Unloading trucks assigned to the Civilian Conservation Corps in Missoula in 1933.

A fleet of stake body trucks in Missoula, Mont., ready to haul CCC workers and supplies wherever they may be needed.

By the late 1930s, mechanics were rapidly replacing blacksmiths and wranglers. This is a scene in the heavy equipment repair shop in 1939 at Missoula, Montana.

The radio station in Missoula, Montana in 1941. This site is now surrounded by homes in the heart of a Missoula subdivision off Whitaker Drive. The building is now owned by the Missoula, Montana Parks and Recreation Department.

All the equipment in this 1936 Forest Service radio station near Missoula, Montana, could be replaced today by a unit that would probably fit into one of the table's desk drawers.

Federal Building

Laying the foundation for the U.S. Post Office and Federal Building in downtown Missoula, Mont., in 1911. The original structure was completed in 1913.

Federal Building in downtown Missoula, Montana, being expanded in 1927.

This east addition was completed in 1928. It became the new Northern Region Headquarters for the U.S. Forest Service in 1936 and still serves this purpose today.

Working in 1936 on the "Forest Service" wing, the last addition to the Federal Building in downtown Missoula.

This structure is on the National Register of Historic Places. The original structure was completed in 1913, the east addition in 1928 and the "Forest Service Wing" in 1936. It has served as Northern Region Headquarters of the U.S. Forest Service since 1913.

FOREST SERVICE IS UNDER SINGLE ROOF

All Branches Brought Into Federal Building; Last Moving Today.

All of the Forest Service offices in Missoula are now under the same roofs in the Federal building.

Movement of the last of the offices from the Higgins block to the Federal building was completed Tuesday.

For years the Federal offices in Missoula were scattered in various buildings, but now most of them have been brought into the Federal building, though the Weather bureau remains in the Montana block, but is soon to move on the third floor of the first Federal building structure, above the postoffice.

The Forest Service is centered in the annex and the east wing, with only C. S. Brothers, assistant to the solicitor, remaining in the old Federal building at the present time.

Forester F. A. Silcox, when here last week, expressed his surprise and pleasure at the size of the Federal building and of the quarters provided for the Forest Service here.

The west part of the front of the Missoula Federal Building in 1931. The lower floor is the downtown Post Office.

The east part of the front of the newly expanded Federal Building in downtown Missoula, Montana, in 1931. The birch tree on the right side of the picture is still there, but much, much larger.

Montana School of Forestry seniors (and the bus in which they traveled), visiting the Priest River Experimental Forest in Idaho. They reviewed research underway in this location.

Speer Hall, the first School of Forestry building at the University of Montana, Missoula. The building was constructed in 1913 and the School was moved to its present site in 1918. This frame building was torn down in 1955.

CHAPTER ELEVEN: Forest Service Personnel

Elers Koch in 1942. This is a man who helped determine what lands belonged in the National Forests of the United States, then dedicated the rest of his life to their development, protection and management—he was a giant in the Northern Region of the Forest Service.

Franklin B. Hough was a physician, historian, statistician and self-taught forester. His was a strong, early voice urging something be done to stop the destruction of forests in the United States. From 1876 to 1883, prior to the National Forests being moved from the Department of the Interior to the Department of Agriculture, he served as federal forestry agent in the Department of Agriculture. This makes him the first Chief of the U.S.D.A. Forest Service.

Frank Fenn (1853-1927) was a very versatile individual. A naval academy graduate, educator, rancher, lawyer, state legislator, army veteran, newspaperman and superintendent of all Forest Reserves in Idaho. In 1908 he was the Supervisor of the Clearwater National Forest and later the Selway National Forest.

Gifford Pinchot, the father of the U.S. Forest Service. He was a tall, dynamic far-sighted man who was the first Chief of the U.S. Forest Service in the Department of Agriculture. Of him, President Theodore Roosevelt said "I believe it is just to say that among the public officials who under my administration rendered literally invaluable service to the people of the United States, he, on the whole, stood first." Trained in Europe, he was a National leader in advocating conservation. He strongly believed in the development of natural resources, and that they be used wisely but not exploited. His philosophies are applicable in the management of today's National Forests.

Chronology of Northern Regional Foresters

William P. Greeley—1908-1911
Chief—1920-1928
Ferdinand A. Silcox—1911-1917
Chief—1933-1939
Richard H. Rutledge—1917-1920
Fred Morrell—1920-1929
Evan W. Kelley—1929-1944
P. D. Hanson—1944-1956

Were the early Rangers as rugged as the land they administered? You be the judge. This is a Pre-fire Ranger's Meeting in the old Bitterroot-Idaho National Forest. Standing in the rear: Rackliff, Litchfield, Clark, Squires, Kelly, Squires, Brundige, L. Fitting, Thenon, Harrington, R. Fitting, Kirchner. Kneeling, middle row: Durant, Cressler, Garrison, Brown, Trenary, Gross, Porter, "Mascot." Seated front row: Cox, Crossley, Forest Supervisor Frank Fenn, Chamberlin, McPherson.

Meeting of Forest Supervisors in Helena, Montana, not long after the U.S. Forest Service was created.

The "Bunch"– Selway Ranger Meeting in April, 1911.

At the Supervisor's Office on the old Madison National Forest in 1911. The occasion for the meeting is thought to be a Ranger's Meeting.

Members of the 1913 Regional Office staff.

Caricatures and "in" comments indicate how close the U.S. Forest Service officers were to the public they served. They were not a nameless, shapeless, bureaucracy.

Forest Supervisor's meeting in Missoula at the Federal Building in 1917. Perhaps the suits, hats and ties are veneer for a rugged body that is more familiar on the back of a horse and following a trail through the mountains.

Gifford Pinchot visiting Forest Service officers at the Ninemile Ranger Station in 1937. Major Kelley is to his right.

A 1939 field trip to the Lick Creek Timber Sale in the Bitterroot National Forest. This early Forest Service timber sale was personally reviewed on-the-ground and approved by Gifford Pinchot.

William P. Greeley, Northern Regional Forester 1908-1911.

F.A. Silcox, Northern Regional Forester 1911-1917.

Richard H. Rutledge, Northern Regional Forester 1917-1920.

Fred Morrell, Northern Regional Forester 1920-1929

Major Evan W. Kelley, Northern Regional Forester 1929-1944.

P.D. Hanson, Northern Regional Forester 1944-1956.

Major Evan W. Kelley

Major Evan W. Kelley was Regional Forester of the Northern Region for 15 years, from 1929 until his retirement in 1944. Born in October, 1882, Kelley began work in the California gold fields when he was 14 years old. He was not a forester or a college graduate. He only had a common school education. (He was, however given the degree of Master of Forest Engineering in 1940 by the University of Montana). He joined the U.S. Forest Service in 1906 and worked his way up through the various grades.

His first job was as a "Forest Guard" on the Yuba National Forest in California. His pay was $60 a month and he had to furnish his own horse, pack animals and tools. He could have made more in the gold fields, but he said that he had "a conscious interest in better treatment of the forests of the country." Prior to World War I, he served as an Assistant Ranger and Ranger on the Tahoe Reserve, Forest Supervisor of the Eldorado National Forest and Regional National Forest Examiner in charge of general forest improvement activities in the California Region.

During Word War I he joined the army and went overseas as a captain in the Tenth Engineers, a forestry regiment, and came home with the rank of major. He was then known as "Major" Kelley the rest of his life. After the war, he returned to work for the Forest Service and in 1920 transferred to the Chief's Office in Washington D.C. as an inspector. He served in this position until his appointment as Regional Forester for the Northern Region. In total, Kelley served the Forest Service for 46 years.

In summing up his career, Kelley wrote, "If I had my life to live again I would choose the same course but its details would be altered at many points."

Avery District Ranger on a trail in the St. Joe National Forest, Idaho in 1920.

Greeley, Tanner, Mason, Silcox, Gerard, Wolff, Graves and Stickney in the early days at the Priest River Experiment Station in Idaho. These were prominent men who played important roles in the establishment of the Forest Service.

Forest Service work camp in 1903, Kaniksu National Forest, Idaho.

The Forest Supervisor's Office at Wallace, Idaho in 1908. Supervisor R.H. Rutledge is dictating to Forest Guard Knox McAnalley.

The 1910 Ranger Meeting at Point Of Rocks Ranger Station near Olney, Montana. Standing left to right: Brunner, Leibig, Cosley, Christenson, Wise, Woesner, Eastland, Taylor and Asst. Regional Forester Silcox. Sitting right to left: Haines, Bissell, DeGroot, Clarke, Owens, Heirick.

Ranger's wives would occasionally accompany their husbands. Mrs. Paul Wohlen is shown here in 1917 on Rocky Ridge, Lochsa Ranger District in the Clearwater National Forest.

A Ranger finds a high point to rest in the 1920s.

A typical ranger of the 1920s.

Attendees of a Ranger Training School held at the Priest River Experimental Forest, Idaho in 1926.

A Ranger on skis in the 1920s. The pole is used for balancing as well as pushing.

Rangers McGregor and Edwards share a joke at the West Yellowstone Ranger Station office in 1936.

On game study work in 1937.

Rangers Mendenhall, Thol and Hutchinson having lunch at Meadow Creek in 1927, Flathead National Forest.

Planting fish brought in by pack train in 1939.

Trail construction in the Kootenai National Forest in 1921.

Interned Italian Nationals

Fort Missoula is located on the southern edge of Missoula. From 1933 to 1942 it served as a regional headquarters for CCC camps in the Northern Region and a supply center for camps in the Northwest.

In 1941 with war raging in Europe the fort was taken over by the Immigration and Naturalization Service to house interned Italian seamen and other nationals who were stranded at American seaports or on American territory.

About 1,200 Italians were incarcerated here until December 1943. Many of these men were hired out to local ranger districts to relieve the shortage of manpower caused by America's entry into the war in December 1941.

Italian seamen leaving Brown's Creek Camp to work on the Musselshell Road near Pierce, Idaho.

Italian seamen in 1943 occupied this Brown's Creek Camp on the Lochsa Ranger District, Clearwater National Forest.

Italian seamen cooking at a camp on the Coeur d'Alene National Forest.

An Italian prisoner of war work detail on the Coeur d'Alene National Forest.

A ponderosa pine display used during American Forest Week in 1926 in Missoula, Mont. This section of the log is 670 years old.

H.T. Gisborne of the Northern Rocky Mountain Forest Experiment Station lecturing to University of Montana Forestry School seniors in 1940 on fire weather studies at the clear-cut inflammability station on the Priest River Experimental Forest.

Relief workers lined up in Spokane, Wash. in 1936 for Blister Rust Control work.

Men using a horse grader on the Little North Fork trail in 1925.

Relief workers assigned transportation in Spokane, Wash., in 1936 for Blister Control work.

Establishment and Modification of National Forest Boundaries

From the establishment of the Forest Reserves in 1891 and the National Forests in 1905, there have been many changes in boundaries of the various units as well as addition and deletion of lands within the boundaries. Some Forest Reserves became National Forests and their name still survives. Other Forest Reserves and National Forests were created and later consolidated with other units or just eliminated as being lands not suited for National Forest purposes.

The basic rationale for most land adjustments was management efficiency. As technology, transportation and communication systems improve, it becomes practical to have larger Ranger Districts. National Forests can also be consolidated when there will not be a loss in management quality and there would be an increase in efficiency.

PRESENT NATIONAL FORESTS IN THE NORTHERN REGION & SUPERVISOR'S OFFICE	ESTABLISHED	STATE
Beaverhead, Dillon, Mont.	July 1, 1908	Mont.
Bitterroot, Hamilton, Mont.	Feb. 22, 1897	Mont. & Idaho
Clearwater, Orofino, Idaho	July 1, 1908	Idaho
Custer, Billings, Mont.	July 1, 1908	Mont. & N.Dak.
Deerlodge, Butte, Mont	July 1, 1908	Mont.
Flathead, Kalispell, Mont.	Feb. 22, 1897	Mont.
Gallatin, Bozeman, Mont.	Feb. 10, 1899	Mont.
Helena, Helena, Mont.	April 2, 1906	Mont.
*Idaho Panhandle National Forests,		
Coeur d'Alene	Nov. 6, 1906	Idaho
Kaniksu	June 26, 1908	Idaho & Wash.
St. Joe	July 1, 1911	Idaho
Kootenai, Libby, Mont.	Aug. 13, 1906	Mont.
Lewis and Clark, Great Falls, Mont.	June 9, 1903	Mont.
Lolo, Missoula, Mont.	Sept. 20, 1906	Mont.
Nez Perce, Grangeville, Idaho	July 1, 1908	Idaho

*The Idaho Panhandle National Forests are the Coeur d'Alene, Kaniksu and St. Joe all administered by one Forest Supervisor whose headquarters is in Coeur d'Alene, Idaho.

NATIONAL FORESTS OR RESERVES DELETED FROM THE NORTHERN REGION

Priest River	Hell Gate	Colville
Jefferson Little Belt	Long Pine	Weiser
Long Pine Madison	Ekalaka	Cabinet
Short Pine	Yellowstone	Snowy Mtn.
Otter	Slim Buttes	Absaroka
Big Hole	Blackfeet	Cave Hills
Little Belt Mtns.	Pryor Mtn.	Pend Oreille
Sioux	Highwood Mtns.	Missoula
Beartooth	Dakota	Elkhorn
Little Rokies	Crazy Mtns.	Selway
Big Belt	Palouse	Jefferson

Bibliography

Benson, Jane Reed, *Thirty-two Years In The Mule Business*, Missoula, Montana: U.S. Dept. of Agriculture, Forest Service, Lolo National Forest, 1980.

Clark, Dr. Wilson F., *Custer National Forest Lands—A Brief History*, Billings, Montana: U.S. Dept.of Agriculture, Forest Service, Custer National Forest.

Cochrell, Albert N., *A History Of The Nezperce National Forest*, Missoula, Montana: U.S. Department of Agriculture, Forest Service, revised 1970.

Cohen, Stan, *A Pictorial History Of Smokejumping*, Missoula, Montana: Pictorial Histories Publishing Co. Inc. 1983.

Cohen, Stan and Miller, Don, *The Big Burn, The Northwest's Fire Of 1910*, Missoula, Montana: Pictorial Histories Publishing Co. Inc. 1978.

Cooley, Earl, *Trimotor and Trail*, Missoula, Montata: Mountain Press Publishing Co., 1984.

Doetch, Bettie, *History Of The Kootenai National Forest*, Libby, Montana: U.S. Dept. of Agriculture, Forest Service, Kootenai National Forest, 1976.

Gildart, Robert C. (Bert), *Montana's Early-Day Rangers*, Helena, Montana: Montana Magazine Inc. 1985.

Guth, A. Richard and Cohen, Stan, *Red Skies of '88*, Missoula, Montana: Pictorial Histories Publishing Co. Inc. 1989.

Hartig, Louis, *The Story Of A Ranger District And Its People In The Clearwater National Forest*, Dubuque, Iowa: Kendall Hunt Publishing Co. 1989.

Kresek, Ray, *Fire Lookouts Of The Northwest*, Fairfield, Washington: YE Galleon Press 1984.

Smith, Steve, *Fly the Biggest Piece Back*, Pictorial Histories Publishing Co., Inc., 1988.

Space, Ralph S. *The Clearwater Story*, Missoula, Montana: U.S. Dept. of Agriculture, Forest Service 1964.

________ *TIMBER*, Historical Museum at Fort Missoula, Missoula, Montana: 1990.

U.S. Dept. of Agriculture, Forest Service *Early Days In The Forest Service*, Volumes 1 through 4, Missoula, Montana: Office of the Northern Regional Forester.

U.S. Dept of Agriculture, Forest Service *Establishment And Modification Of National Forest Boundaries*, Washington, D.C. 1985.

U.S. Dept. of Agriculture. *The Principal Laws Relating To Forest Service Activities*, Agriculture Handbook 453, U.S. Government Printing Office—revised September 1953.

U.S. Dept. of Agriculture. *The Use Of The National Forests*, Issued June 14, 1907.

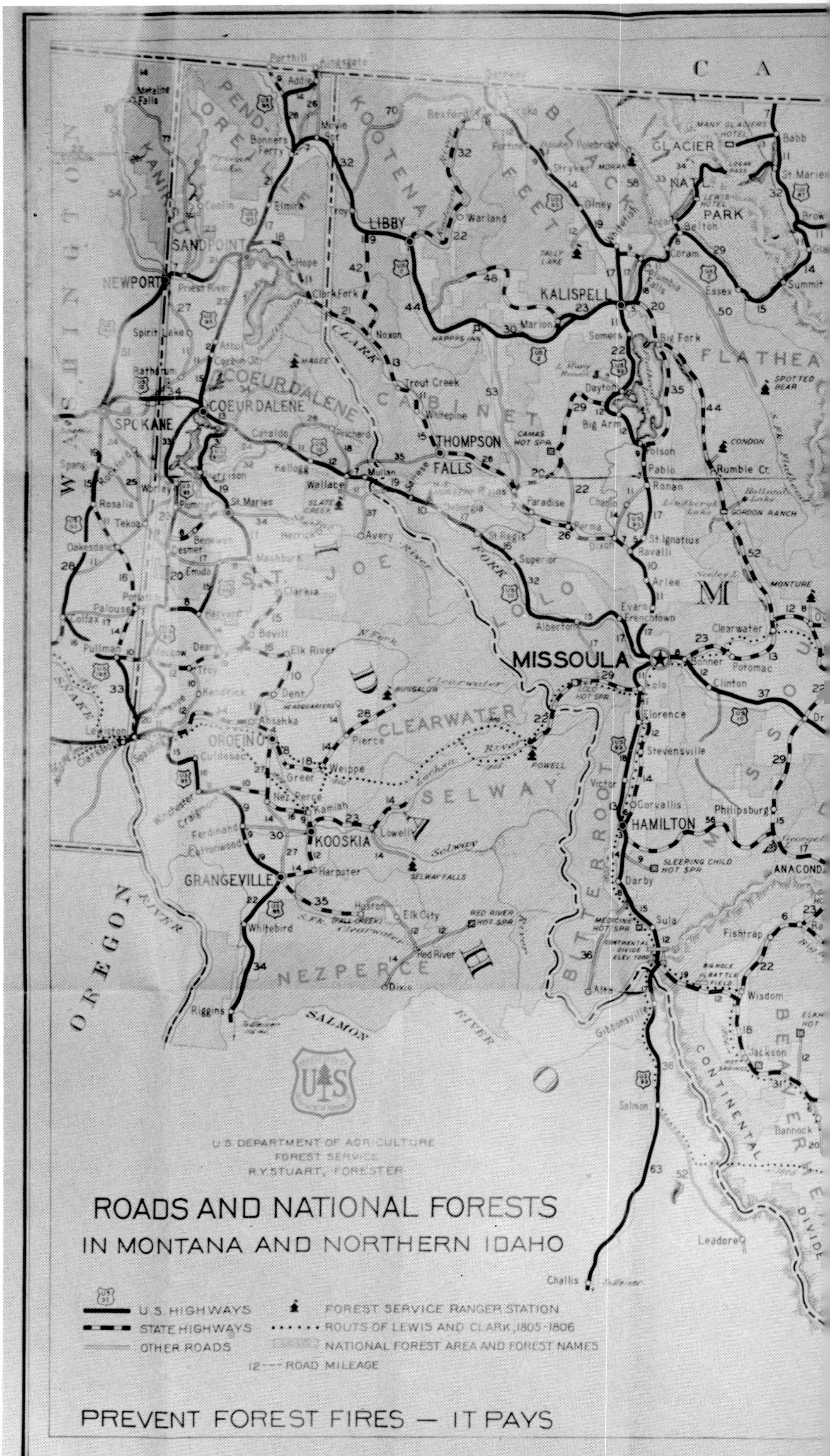
U.S. DEPARTMENT OF AGRICULTURE
FOREST SERVICE
R.Y. STUART, FORESTER
ROADS AND NATIONAL FORESTS
IN MONTANA AND NORTHERN IDAHO
U.S. HIGHWAYS
STATE HIGHWAYS
OTHER ROADS
FOREST SERVICE RANGER STATION
ROUTS OF LEWIS AND CLARK, 1805-1806
NATIONAL FOREST AREA AND FOREST NAMES
12--- ROAD MILEAGE
PREVENT FOREST FIRES — IT PAYS
MISSOULA
KALISPELL
HAMILTON
GRANGEVILLE
SPOKANE
COEURDALENE
SANDPOINT
NEWPORT
LIBBY
THOMPSON FALLS
KOOSKIA
OROFINO
GLACIER NAT'L PARK

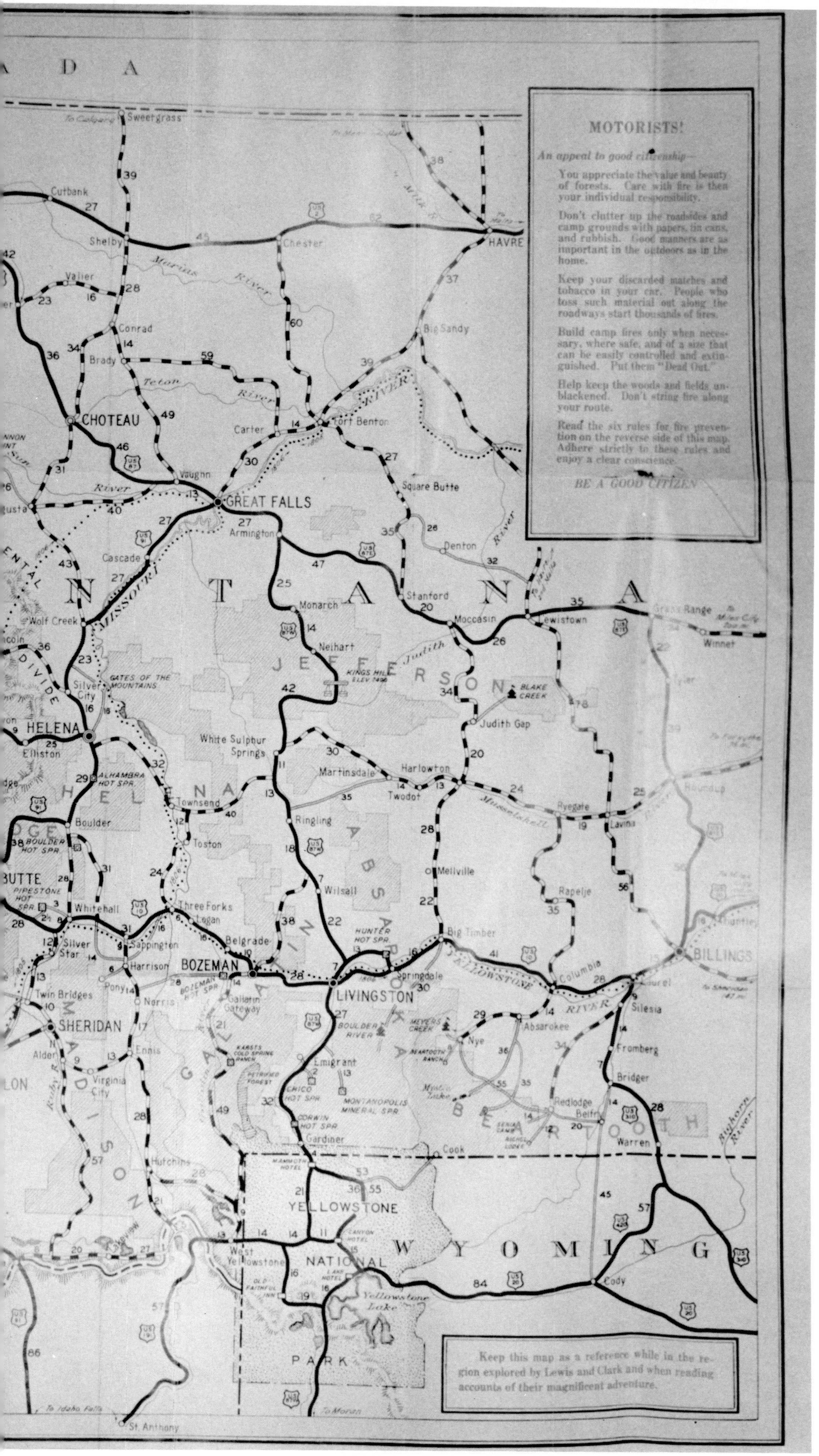
MOTORISTS!
An appeal to good citizenship—
You appreciate the value and beauty of forests. Care with fire is then your individual responsibility.
Don't clutter up the roadsides and camp grounds with papers, tin cans, and rubbish. Good manners are as important in the outdoors as in the home.
Keep your discarded matches and tobacco in your car. People who toss such material out along the roadways start thousands of fires.
Build camp fires only when necessary, where safe, and of a size that can be easily controlled and extinguished. Put them "Dead Out."
Help keep the woods and fields unblackened. Don't string fire along your route.
Read the six rules for fire prevention on the reverse side of this map. Adhere strictly to these rules and enjoy a clear conscience.
BE A GOOD CITIZEN
Keep this map as a reference while in the region explored by Lewis and Clark and when reading accounts of their magnificent adventure.
Sweetgrass
Cutbank
Shelby
Chester
HAVRE
Valier
Conrad
Brady
Big Sandy
CHOTEAU
Carter
Fort Benton
Vaughn
GREAT FALLS
Square Butte
Armington
Denton
Cascade
Stanford
Monarch
Moccasin
Lewistown
Grass Range
Winnet
Wolf Creek
Neihart
JEFFERSON
Silver City
GATES OF THE MOUNTAINS
KINGS HILL
BLAKE CREEK
Judith Gap
HELENA
Elliston
White Sulphur Springs
Martinsdale
Harlowton
Twodot
Ryegate
Lavina
Roundup
ALHAMBRA HOT SPR.
Townsend
Toston
Ringling
Boulder
BOULDER HOT SPR.
Mellville
Rapelje
BUTTE
PIPESTONE HOT SPR.
Whitehall
Three Forks
Logan
Wilsall
Big Timber
Belgrade
HUNTER HOT SPR.
BILLINGS
Silver Star
Sappington
Harrison
BOZEMAN
Springdale
Columbia
Laurel
LIVINGSTON
Twin Bridges
Pony
Norris
BOZEMAN HOT SPR.
Gallatin Gateway
BOULDER RIVER
MEYERS CREEK
Absarokee
Silesia
SHERIDAN
Ennis
Nye
Fromberg
Alder
Emigrant
Bridger
Virginia City
CHICO HOT SPR.
MONTANAPOLIS MINERAL SPR.
Redlodge
Belfry
CORWIN HOT SPR.
Gardiner
Warren
Cook
Hutchins
MAMMOTH HOTEL
YELLOWSTONE
West Yellowstone
NATIONAL
PARK
OLD FAITHFUL INN
Yellowstone Lake
W Y O M I N G
Cody
St. Anthony
MISSOURI
YELLOWSTONE RIVER
BEARTOOTH
MADISON
GALLATIN

About the Authors

A. RICHARD GUTH spent most of his career as a forester employed by the U.S. Forest Service. He retired in 1986 after serving more than 30 years.

During the last 12 years of his career with the Forest Service Guth served as Public Information Officer in the Northern Regional Headquarters. His responsibilities during this period included the Region's History Program as well as extensive writing and editing assignments. During previous assignments he served as an Assistant District Ranger, District Ranger and National Forest Staff Officer with responsibilities in timber, wildlife, recreation, and watershed management.

A 1951 graduate of the University of Michigan, Guth has taken over 30 hours of post graduate work at the University of Montana in Journalism, Interpersonal Communications and Forestry. He has also taken additional course work at in Communications at the University of Idaho. Since his retirement, Guth has remained active as a professional forester serving as a volunteer with the Northern Regional Information Office. He also serves as editor of the Society of American Foresters newsletter for the State of Montana and editor of the newsletter for the Northern Rocky Mountain Retirees Association. He is also president of Guth Writing & Natural Resource Consulting, Inc. With Stan Cohen, Guth co-authored "Red Skies of '88" in 1989. He has also authored several articles which appeared in national publications.

STAN COHEN is a graduate geologist from West Virginia University. His Forest Service career includes working with a fire crew in Oregon and California, three summers at the Intermountain Fire Sciences Laboratory in Missoula and serving on the geologic staff in the Northern and Alaska Regions. He has been in the publishing business the past 14 years. He has also authored or co-authored 37 books and published well over 100.